Copyright © 1983 by Karl Fulves.
All rights reserved under Pan American and International
Copyright Conventions.

Published in Canada by General Publishing Company,
Ltd., 30 Lesmill Road, Don Mills, Toronto, Ontario.
Published in the United Kingdom by Constable and Com-
pany, Ltd.

Self-Working Number Magic: 101 Foolproof Tricks is a new
work, first published by Dover Publications, Inc., in 1983.

Manufactured in the United States of America
Dover Publications, Inc., 180 Varick Street, New York,
N.Y. 10014

Library of Congress Cataloging in Publication Data

Fulves, Karl.
    Self-working number magic.

    1. Mathematical recreations.   I. Title.
QA95.F85   1982   793.7′4   82-9453
ISBN 0-486-24391-5   AACR2

INTRODUCTION

Number tricks can be performed at any time because most often all you need to do them is a pad and a pencil. Tied in with patter about numerology, even the simplest tricks take on an air of profound mystery. So, with pencil and paper in hand and suitable presentation in mind, you can demonstrate an assortment of baffling tricks with numbers.

Many number tricks in this book are based on cleverly concealed mathematical principles, but the mathematics will not be discussed here. The accent here is on streamlined tricks that are easy to grasp and to perform.

The chapters on Lightning Calculators, Giant Memory, and Magic Squares represent some of the best current thinking on these subjects. The material sidesteps mathematical sophistication in favor of practical methods that can be learned quickly and put into operation almost immediately.

Numbers pervade our daily lives. There are telephone numbers and zip codes, Social Security numbers and driver's license numbers. Almost everyone believes that some numbers are luckier than others. In presenting the tricks in this book, try to connect the routines with numbers that have meaning to the spectator. Thus, instead of having a spectator jot down a random number, have him jot down the last two digits of his telephone number or the first two digits of his license plate number. He will then think that those digits are somehow important to the outcome, and this is the real secret of success with material of this kind.

Describing a number trick some years ago, an authority on the subject wrote, "Like most mathemagical stunts, this one appears complicated, but is really quite simple." My belief is that number tricks should not appear complicated. If arithmetic is involved, it should be concealed or made to seem an incidental part of the trick. The routines, stunts, and betting games in this book were assembled with this point in mind—to put before the reader a collection of easy but baffling number tricks.

For their assistance in gathering the material described here I would like to thank Charles D. Rose, Sam Schwartz, Fr. Cyprian, Howard Wurst, Joseph K. Schmidt, and the major contributor, Martin Gardner.

KARL FULVES

CONTENTS

SELF-WORKING
NUMBER MAGIC
101 FOOLPROOF TRICKS

AMAZING NUMBER MAGIC

The tricks, stunts, and games described in this chapter provide clear plot ideas to the spectator. But although they are seemingly simple in effect, they lead to surprising conclusions. Although based on established principles, they are generally not well known, even to magicians, so they are likely to fool even the most sophisticated spectator.

No. 10 in this chapter is an excellent mystery called "Lightning Strikes Thrice." Based on Paul Swinford's description of a trick of Stewart Judah, it produces a baffling coincidence effect by means of ingenious subterfuge. Follow it with a trick like "Stunumbers" in the next chapter and you will have an exceptional combination.

1 THE NINE MYSTERY

"I just found something I had written in grade school," the magician says. He holds up the piece of paper shown in Figure 1. Written on it in script is "nine." The magician continues, "I remember it clearly. We were asked to write our age on a piece of paper. I wrote this number and handed in the paper. Too late I realized that I hadn't dotted the 'i.' The teacher had the paper so there was no way to correct the mistake."

The magician turns the paper over and places it writing-side down on the table.

Then he picks up another piece of paper and writes or draws the dot that's supposed to be on the "i." He says, "This was when I performed my first magic trick. I wrote a dot like this, then I erased it by magic." The magician rubs the paper and the dot mysteriously vanishes.

Picking up the "nine" paper, he turns it over. Now there is a dot over the "i," as shown in Figure 2. "Needless to say," he concludes, "I got an A + for my efforts."

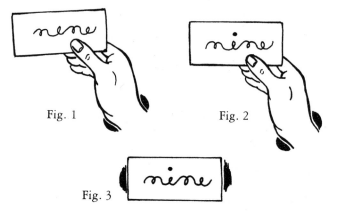

Fig. 1   Fig. 2

Fig. 3

METHOD: This delightful trick is based on an idea of John Hamilton's. The secret is shown in Figure 3. The number nine is written in script in such a manner that it reads the same right side up or upside down. The dot is placed as shown.

When you hold the paper as in Figure 1 to display it, the right thumb covers the dot. Place the paper writing-side down on the table. Then pick up another piece of paper and pretend to draw a dot or circle. The spectators hear you write, but what they hear is the sound of the fingernail scraping against the paper.

Place the pencil aside. Pretend to cover the dot with the fingers. Rub the fingers back and forth. The dot seems to have vanished. Then pick up the "nine" paper, but make sure it is turned around so that the dot is uppermost. Now the number appears as shown in Figure 2. The dot has been magically transferred from one piece of paper to another.

2 COMPUTER DATING

Some computer dating systems use astrology and numerology to bring compatible people together. When you meet a happily married couple, offer to demonstrate how numerology applies to computer matching.

The husband and wife sit across the table from one another. Give each a piece of paper. Ask each to jot down a single-digit number, that is, any number from 1 to 9. You are standing or sitting at the husband's side of the table so you can't see his wife's number.

Ask her to do the computer figuring as follows. She is to double her number, add 2, multiply the result by 5, and subtract 3. Naturally the computer would do this instantly. When she has a result, ask her to hide it so you can't see it.

"The computer would have your number on file," the magician says to the wife. "And after doing that bit of figuring it would look for a compatible number in its memory bank. What result did you get?"

She might say 27. "You can see how well it works. You must be very compatible because you wrote a 2 and your husband wrote a 7." And they did!

METHOD: The puzzling aspect of this trick is that you never know the wife's number. It appears as if the result is completely random, yet the outcome is always a number that matches both the husband's and the wife's number.

The secret is that you have only to know the husband's number for the trick to work. Whatever digit he chose, subtract it from 10. In this example he wrote 7. Subtract it from 10 to get 3. In this case 3 is the key number.

Have the wife double her number, add 2, and multiply the result by 5. Then you have her subtract the key number, which in our example is 3. When she does, she will get back *both* of the original numbers.

To take the above example, she might write 2 and her husband 7. She would double 2 to get 4, add 2 to get 6, multiply the result by 5 to get 30, and subtract the key number 3 to get 27.

Remember that the key number is the result you get when you subtract the husband's number from 10. It is not necessary to make it obvious that you know his number. You can glimpse it and then turn away. All that is required is a quick glimpse of the number for the work to be done.

As simple as this trick is, it has a strong effect. The happy couple will be delighted to discover that the computer verified their compatibility and they will be mystified as to how the computer knew.

3   RESISTIBLE RITHMETIC

"Some numbers refuse to be added," the magician says. "For example, if I ask you to write down the number 81, and under it I write the number 10, when these numbers are added they will *not* total 91."

The spectator is handed a pad and pencil. He writes the number 81, but he does so with the writing side of the pad away from him as shown in Figure 4. Then the magician takes the pad and under the spectator's number he writes 10. The pad is then given to someone else, who totals the numbers and arrives at the sum of 28. The arithmetic is correct, the numbers are not switched, there are no confederates, yet the total is different from the spectator's expectations.

The reason why is even more puzzling than the mysterious sum of 28 because the spectator discovers that although he definitely wrote 81, that is *not* what appears on the pad!

METHOD: The key to it is the manner of writing shown in Figure 4. When the spectator holds the pad as shown and writes the number 81 on the reverse side, the writing reverses itself, changing 81 to 18. You have only to try it once to convince yourself that it works.

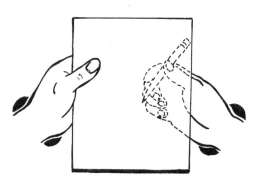

Fig. 4

To present the trick, draw a scribble on the pad, saying that it respresents a mysterious method of hypnotizing people into doing strange arithmetic. So that the spectator won't be influenced by this potent symbol, have him hold the pad with the back to him and jot down the number 81 on the other side.

Take the pad from him and under his number write the number 10. Draw a line below the two numbers and have someone total them. He will get 18 + 10 = 28. The first spectator will want to check the addition, but when he sees the writing he will be surprised to discover that his numbers have rearranged themselves. The hypnotic spell worked.

4 STEP RIGHT UP

You offer the spectator an engagingly easy way to make a quick dollar. Deal 16 slips of paper or business cards in a row on the table. Show him that on half of the slips you have written, "I Owe You $1," and on the other slips you have written, "You Owe Me $1." The slips are writing-side down so no one can see the writing.

Tell the spectator to jot down the numbers 1, 2, 4, 8 on a pad in any order and to put plus and minus signs between the digits. He can use all plus signs or all minus signs or any combination of plus and minus signs. He might write the numbers like this:

$$4 \ 2 \ 8 \ 1$$

And he might decide to use this combination of plus and minus signs:

$$4 + 2 + 8 - 1$$

You've explained beforehand that he's going to total the digits according to the indicated arithmetic operations, and further that he will disregard any negative sign that may show up in the answer, and finally that, whatever the result, he will count to that slip from his left. Remember that all this is explained beforehand so you cannot back out or change the rules.

In the above example he would perform the indicated addition and subtraction operations and arrive at the total of 13. He counts to the 13th slip, counting from left to right, and finds on it the words, "You Owe Me $1," meaning that he has just lost a dollar to you.

The game can be repeated any number of times. The spectator *always* loses.

METHOD: When you place the slips in a row, make sure the "You Owe Me $1" slips are in every odd position. The result is then automatic. You cannot lose because he must end up on an odd-numbered slip. If, for example, he arranges the slips in the order 2, 4, 1, 8, and puts in plus and minus signs as follows:

$$2 + 4 - 1 - 8,$$

he will arrive at an answer of -3. As mentioned, he disregards the minus sign in the answer. He will thus arrive at the slip in the third position from the left and he loses.

You can cause the sum to be an even number by noting that the controlling factor is the digit 1. If you use just the digits 2, 4, 8, in any order, with any combination of plus and minus signs between them, the spectator must arrive at an even number.

To exploit this, write the numbers 1, 2, 4, 8, 16 on separate business cards or blank squares of cardboard. Lightly mark the back of the 1-card so you can recognize it at a glance.

Don't deal the 16 slips of paper in a row. Instead, arrange them in a packet so that the "I Owe You" slips alternate with the "You Owe Me" slips. The top slip of the packet is an "I Owe You" slip.

Hand the spectator the packet of five business cards. Tell him to mix them writing-side down and choose any three or four cards. Simply note whether or not he picks the 1-card. If he does, deal the 16 slips of paper from right to left. If he doesn't pick the 1-card, deal the groups of slips from left to right. Note that the real work is done before the game begins.

Have him arrange the chosen business cards in any order. Then tell him to place any combination of plus and minus signs between the numbers and carry out the arithmetic operations. Whatever the result, he counts to that number, beginning at the left. The word "left" indicates the left end of the row from the performer's view. This point should be made clear to the spectator at the start. You can simplify things by standing at the same side of the table as the spectator. In any event, when he counts to the chosen slip of paper, he will discover that he always ends up owing you money.

5 DOLLAR-BILL POKER

If you want to impress someone with a seemingly incredible talent for total recall, have him remove a dollar bill from his pocket and call out the digits of the serial number to you. Then tell him to circle one digit. After he does this, have him call out the remaining digits in *any* order. As soon as he does, you announce the digit he didn't call out!

METHOD: Oddly enough, if you tried to perform "Dollar-Bill Poker" by memorizing the digits, you would find it a difficult task. The reason is that the second time the spectator calls out the digits, he calls them out in any order. The trick can be done with memorization but there is a sneaky way to do the trick that is much easier.

When the spectator calls out the digits the first time, mentally add them together. Remember the total.

When he calls out the digits the second time he will call seven of the eight digits but will not call out the circled digit. Add the digits he calls out, then subtract this total from the previous total. The result is the circled number.

For example, say he calls out the serial number 48253176. As he calls out the digits, you mentally add them together as follows:

$$4 + 8 + 2 + 5 + 3 + 1 + 7 + 6 = 36$$

Remember the total of 36.

He might decide to circle the digit 3. Then he calls out the remaining digits in any order. When he does, mentally add them. The result may look like this:

$$6 + 7 + 1 + 5 + 2 + 8 + 4 = 33$$

Simply subtract the second total from the first, getting $36 - 33 = 3$. Then announce that the circled digit is a 3.

The trick is simple but you should make it look difficult. When the spectator reads off the digits the first time, you can ask him to read them again. This may be necessary anyway if you didn't have a chance to add all digits, but it will make the trick appear more difficult. The throw-off is in your opening comments that you've been practicing memory tricks and can recall seven-digit and eight-digit numbers with ease. This *seems* like a logical explanation as to how you knew the circled digit, but even a memory expert will have trouble duplicating the trick if he tries to memorize the numbers.

Later in this book, in the chapter on Giant Memory, a system will be given that allows you to actually memorize an eight-digit number at a glance. The trick, called "Serial Secret" (No. 88), allows you to call out the memorized number forward or backward. After you have done "Serial Secret" you may want to switch to "Dollar-Bill Poker." It appears to be an even more impressive feat of memorization, but now you know that a fake method is used. Used together, the two tricks form a strong combination.

6 FAMILIAR SPIRIT

It is well known that each person has a twin, someone who looks the same, talks the same, and even thinks the same. The question is, where on the planet would one find his twin? By means of the International Telepathic Dialing System, one can locate the city where his twin lives.

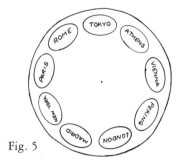

Fig. 5

The Telepathic Dialing System is shown in Figure 5. Place it before a spectator. Study him for a moment and then say, "I'm getting a mental picture of the city where your exact double can be found. He's on the same wavelength as you. In fact, right now he's using the Telepathic Dialing System to determine where *his* double lives."

You jot down the word "Paris" on a piece of paper without anyone seeing it. Fold the paper and place it in a drinking glass or under a cup.

Now tell the spectator to take the last four digits of his home or business phone number and jot them down. Then have him jot down the same four digits in any scrambled order. Tell him to subtract the smaller number from the larger. Direct him then to add together all the digits in the total to arrive at a Telepathic Total.

Ask him to place his finger or the point of a pencil on Rome in the dial and count clockwise, with Rome as 1, Tokyo as 2, Athens as 3, and so on until he has advanced a number equal to his Telepathic Total. Surprisingly enough, he will arrive at Paris, exactly matching your prediction that his double lives in France.

METHOD: Make the circular dial of Figure 5 from cardboard. Nothing else is required. Proceed exactly as described above. If the spectator begins his count on Rome he will inevitably finish on Paris.

One example should make the procedure clear. Say that the last four digits in the spectator's phone number are 1284. He scrambles them and arrives at 8142. Subtracting the smaller number from the larger results in an answer of 6858. Adding together the digits in 6858, we arrive at $6 + 8 + 5 + 8 = 27$. Begin at Rome, counting it as 1, then proceed clockwise 27 spaces. You will finish at Paris.

If someone else wants to try it, place the dial in front of him, but with some other city uppermost. Whatever city is at the top, the city he will end up on will be the one to the left of it in a counterclockwise direction. If, for example, he starts at New York, he will end at Madrid. Thus each spectator will discover his double in a different city.

The trick is puzzling because the spectator can choose any four-digit number. He thus reasons that you could not possibly know what number he will choose, nor how he will scramble the digits. This is true but it has no bearing on the outcome.

When doing the trick, be certain to use a four-digit number based on a *telephone* number. This is not necessary in a mathematical sense but it is important in terms of presentation. The telephone number ties in with the telephone dial and thus seems to be connected to it in some profound way. If you use any random four-digit number, the spectator will quickly conclude that the trick works on a mathematical premise. By making it appear that the spectator's phone number is crucial to the trick, you deepen the mystery.

7 A 5¢ COMPUTER

Using a computer made of paper, you can beat an electronic machine if you know the secret. The paper calculator is shown in Figure 6. It consists of a folder with a window in it and a sliding piece with numbers on it. Place the sliding piece inside the folder as shown in Figure 7 and you have a machine that can beat the most sophisticated electronic calculator.

Have a spectator jot down a three-digit number and then repeat

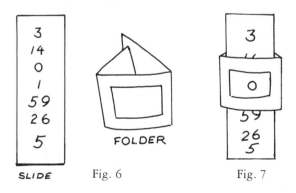

SLIDE   Fig. 6   Fig. 7

it to form a six-digit number. If he jotted down 638, he would repeat this number, arriving at 638,638.

Now ask him to give the slip of paper to another spectator. Emphasize that you do not know the chosen number. Tell this second spectator to divide the given six-digit number by 7 and note the remainder. He can do this with pencil and paper or with a pocket calculator. It makes no difference because you will figure out the remainder *even though you don't know the number.*

The remainder you get with your paper slide rule may be zero, as shown in Figure 7. When the spectator finishes dividing 638,638 by 7 he must admit that the remainder is indeed zero, exactly as your paper computer said.

Whatever answer the spectator got when he divided the number by 7, have him jot this number down on a piece of paper. In our example he would jot down 91,234 (because 638,638 divided by 7 is 91,234). Now have another spectator divide this number by 11 and note the remainder. Once again, even though you don't know the new number, you are able instantly to calculate the remainder when he divides 91,234 by 11.

Whatever answer he got (in our example it would be 8294) have him jot this number down on a slip of paper and hand it to another spectator. Tell this person to divide the new number by 13 and note the remainder. Once again, using just the paper computer, you are able to calculate the exact remainder instantly.

Finally, take the number arrived at by this spectator and jot it down on the sliding portion of the paper computer. You make a few rapid calculations and then show the spectator the number showing through the window. It is his original three-digit number!

METHOD: The trick works itself. All you need do is see to it that zero is the number that shows in the window each time. Don't make it look too obvious. Move the slide back and forth, then slide it so that zero shows.

In our example the spectator writes 638,638. When he divides it by 7 he gets 91,234 with a remainder of zero. When he divides 91,234 by 11 he gets 8294 with a remainder of zero. Finally, when he divides 8294 by 13 he gets 638 with a remainder of zero.

Jot down the last number, 638, on the slide, pretend to make a few calculations, then move the slide so that 638 shows through the window. Show the slide to the first spectator and ask if you correctly calculated his original number. He will be amazed that you arrived at exactly the number he thought of.

8 MARK OF THE BEAST

It is commonly known from the Bible that 666 is the Mark of the Beast, but it is perhaps not as well known that the Beast has a phone number that converts any number to a multiple of 6 or even of 666.

Have the spectator jot down the numbers in his home address. Then have him multiply this number by itself. Then have him multiply it by itself once more. If, say, the given number is 13, he would multiply 13 × 13 × 13 to get 2197.

Then have him subtract his original number. In this example he'd get:

$$\begin{array}{r} 2197 \\ -\ 13 \\ \hline 2184 \end{array}$$

To this you have him add the last four digits of the Beast's telephone number, which, though unlisted, is believed to be 6660:

$$\begin{array}{r} 2184 \\ +\ 6660 \\ \hline 8844 \end{array}$$

The result will always be evenly divisible by 6, the Beast's favorite number.

METHOD: This trick works itself and is valid for any number. Simply have the number multiplied by itself, then multiplied by itself again. The spectator subtracts the given number and adds 6660. The result will be evenly divisible by 6.

If a pocket calculator is handy to make the work easier, you can convert the given number to a number that is always evenly divisible by 666. After the spectator multiplies the given number by itself twice and subtracts the given number, have him multiply the result by 111. Whatever number he gets, it will be evenly divisible by 666.

9 THINK OF A NUMBER

This is a twist on a stunt credited to the poet Samuel Taylor Coleridge. Tell the spectator to think of any number, double it, add

8, take half the result, and then subtract the original number. Finally, have him add the value of the change in his pocket to this number and tell you the result. Immediately you announce how much change he has in his pocket!

The trick can be done while you have your back turned, or while you are out of the room, but the best presentation is to call a friend on the phone, have him go through the calculations given above, and then announce to him how much change he has. Done long distance over the phone, the trick appears impossible.

METHOD: When the spectator announces the result, subtract 4 from it. The number that remains is the value of the change in his pocket.

Say he thinks of the number 15. He doubles it, getting 30, then adds 8 to get 38, then takes half of this number to arrive at 19. He subtracts the original number to get $19 - 15 = 4$. Unknown to him, he will always arrive at 4. If he has 37 cents in change, he adds $37 + 4$ to get 41. When he tells you his result is 41, subtract 4 to determine the amount of change in his pocket.

10 LIGHTNING STRIKES THRICE

A feature of Stewart Judah's impromptu work, this fine trick uses simple props to produce an uncanny coincidence. Judah suggested the use of blank business cards, but three slips of paper will do as well. You are seated at a table with three friends. The trick then unfolds as follows.

"We're going to try a little experiment here, a demonstration of something you've probably heard about, that lightning never strikes twice in the same place. I'm sure you would also agree that lightning never strikes three times in the same place. We're going to check on that to see if it's really true."

Tear three slips of paper from the pad and tear them in half across the middle. Place them on the table as shown in Figure 8. Pieces A and F are the two halves of the same slip of paper. B and E are the two halves of another slip. C and D are the two halves of the third slip. At this point the slips are blank, so disregard the numbers for a moment.

The pieces are placed on the table in seemingly random order, but make sure they are arranged as described above. As you tear them, say, "I usually use six slips of paper, but just to economize

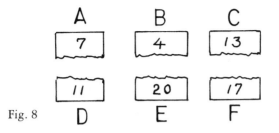

Fig. 8

I'll use three slips and tear them in half. This will gives us six pieces."

Say to the spectator, "I'd like you to give me a number between 1 and 20. Take your time in deciding and don't let me influence your choice. Change your mind as often as you like, but finally decide on one number. It can be the age of one of the people in your family, or how many pounds you'd like to lose this week on your diet, or your lucky number." Eventually the spectator will name a number. We'll assume it is 7. Write the number 7 on slip A as shown in Figure 8.

Ask a second spectator for another number between 1 and 20. He might call out 4. Write a 4 on Slip B as in Figure 8. Finally ask a third spectator for yet another number between 1 and 20. If he says 13, write 13 on slip C as in Figure 8. Each number should have some personal significance to the individual spectator.

Say, "I'm going to write numbers on the papers in the bottom row. I want them to be obviously different from the numbers you chose. On the other hand, I don't want you to think that I've secretly chosen certain numbers beforehand, so I'll do it this way. We'll add the first spectator's 7 and the second spectator's 4 to arrive at 11." Write 11 on slip D.

"Now we'll add the first spectator's 7 and the third spectator's 13 to arrive at yet another random number, in this case 20." Write 20 on slip E.

"Finally, we'll add the second spectator's 4 and the third spectator's 13 to arrive at the random number 17." Write 17 on F. The situation now will look exactly like Figure 8.

Turn all six pieces over so they are writing-side down. Turn your head to one side and have someone mix the pieces around on the table. After this is done, turn back and say, "Six pieces of paper, six random numbers. would you point to one?" Address this question to the first spectator.

When he points to a piece, note how it is torn. With just a glance you can spot the matching portion to this piece. Take the piece he indicated and slide it toward him. Then slide the matching piece toward him.

Have a second spectator indicate one of the remaining pieces. Slide it toward him. Then slide the matching piece toward him.

The third spectator gets the remaining two pieces. At this point one spectator will have pieces A and F, another will have B and E, and another will have C and D.

Turn to the first spectator and have him total his two numbers. In our example the total would be 24. Say, "Let's see if lightning can strike twice. Would you total your two numbers?" This question is addressed to the second spectator.

He totals his numbers and finds that they also total 24. "That's intriguing—24 again, the same total. Now we'll see if the impossible can happen, if lightning can strike not twice but thrice." Turn to the third spectator. Have him turn his two pieces of paper over and total them. The trick is brought to a dramatic finish when it is discovered that his total is also 24. You end with, "Isn't it amazing that this one time we found that lightning can strike thrice? You chose random numbers, yet it all worked out beautifully."

In terms of method or secret there is nothing more to it than described above. The total will be different each time you try the trick, yet the "coincidence" works infallibly.

11   BLUE LIGHTNING

The reason "Lightning Strikes Thrice" works out is that each spectator adds the other two numbers to his own number. In our example the sum of the three original numbers is 7 + 4 + 13, or 24, so that will be sum each spectator is going to arrive at. If you check the sums of Figure 8 you will find that the top three numbers total 24, while the bottom three numbers total twice as much, or 48. All of this information can be used in other ways. An application to a card trick is given here. Needed are two shuffled decks, one blue-backed and one red-backed.

Spectator A removes fewer than nine cards from the top of the red-backed deck. Spectator B and Spectator C remove fewer than nine cards also. Count A's cards and then count B's cards. If A has 7 and B has 4, add them together to get 11. Count 11 cards off the top of the red-backed deck and place it at D, using the position references given in Figure 8.

Now add together B's cards and C's cards. If B has 4 cards and C has 8, the total is 12. Deal 12 cards off the top of the deck and place this packet at F.

To get the third random number, add together C's number and A's number. In the example we would get 8 + 7 = 15. Deal off 15 cards and place them at E. The result thus far is shown in Figure 9.

Fig. 9

Combine A with F and give it to the first spectator. Combine B with E and give the packet to the second spectator. Combine C with D and give this packet to the third spectator. Each spectator now holds the same number of cards, though this is not at all obvious.

Remove a blue-backed deck from its case. You can have this deck shuffled before you proceed. Have each spectator secretly note the card at the same number from the top as the number of cards he holds. If the spectator holds 19 cards, he would note the card 19th from the top of the blue-backed deck.

After each spectator has noted a card in the blue-backed deck, take the deck. Since you know the total of the cards originally dealt at A, B, and C, you know the location of the chosen card in the blue-backed deck. In our example you dealt 7 + 4 + 8 cards at A, B, C, and the total is 19. Beginning at the top of the blue-backed deck, count down to the 19th card. Remove this card plus two others. Hold all three cards in a fan, then ask each spectator if he sees his card. Each will say yes.

Remove the force card, the one that was originally 19th from the top, and place it face up on the table. Discard the other two cards of the fan. Point to the face-up card and say, "Do each of you still see your card?" They will be astounded to discover that each chose the same card.

DEVILISH DATES

Throughout this book an attempt is made to get away from number tricks as exercises in mathematics. The general approach is to make the trick seem important to the spectator by tying it in with something unique to his own life. In the present chapter we deal with significant dates in a person's life, including his age.

A trick like "Mysterious Math" (No. 14) introduces a principle used to determine someone's age, but then it is enlarged upon in "Astrology Cards" (No. 15) to allow you to reveal not only how old a person is, but on which day of which month he was born. The final trick in the chapter, "Stunumbers" (No. 17), is a spectacular example of a number trick where the mathematics is completely concealed.

12   GUESS YOUR AGE?

There are many methods of determining a person's age using simple calculation but the following method is startling in its simplicity and will fool those who know other methods.

Ask the spectator to jot down his age. Then remark that since there are 7 days in a week and 52 weeks in a year, he is to add the number 752 to his age.

When he's done this, state that since there are 12 months in a year, he is to add 12 to the number he got.

Then have him tell you the last digit of his total. He does and you immediately tell him his age!

METHOD: The secret depends in part on calculation but it also depends in part on observation. Since the method is not entirely mathematical, you arrive at the right answer with seemingly too little information. Whatever digit the spectator gives you, subtract 4 from it. If the number is smaller than 4, you must add 10 to it before you subtract the 4.

Suppose he tells you the last digit of his total is 7. Subtract 4, obtaining 3. If he told you the last digit was 3, add 10, getting 13, then subtract 4, arriving at 9. In either case this digit is the last digit of his age.

The rest has nothing to do with mathematics. Simply estimate his age by looking at him. It is reasonably easy to tell if someone is in his 20s, 30s, 40s, etc. If the answer you got from your simple calculation was 3 and you estimate the spectator is in his 20s, announce that his age is 23.

If you find it absolutely impossible to determine whether a spectator is, say, in his 20s or 30s, choose the lower figure. Even if you are wrong the spectator will be flattered that you thought him younger than he really is.

13 FOURTH-DIMENSIONAL DATES

When you meet a happily married couple you have never seen before, you can perform a knockout effect using time telepathy. Hand a pad and pencil to the wife and ask her to jot down the year she and her husband were married. Then have the husband jot down on the pad the year he met his wife. Finally have the wife jot down how many years she and her husband have been married, including the present year. She then gives the pad to her husband, who totals the dates and announces the total.

You say, "Let me see if I can use a bit of fourth-dimensional thinking. You know that time is the fourth dimension. If I mentally go back in time, I can see that, let's see, it's getting clearer, I can't tell when you were married. That part is indistinct. But I can see you two the day you met. It was in 1970, wasn't it?"

Strangely enough, you are right!

METHOD: This comes down to an extremely simple system, so simple that you should disguise it with patter about time telepathy. The idea is to distract the audience with verbal misdirection, and any reasonably serious presentation along the lines of telepathy and the fourth dimension should do the trick.

The secret is this. If I ask you to add together the year of your birth and your age this year when you celebrate your birthday, those two numbers *must* add up to the present year. For example, if this year is 1982, and you were born in 1962, you will be 20 when you celebrate your birthday this year. If we add the year of your birth and your present age, 1962 + 20, we must arrive at the year 1982.

Done this way, the trick is obvious and easily spotted. But you can disguise it by asking the lady to jot down the year she was married, them immediately ask the gentleman for a completely unrelated date, the year he and his wife first met. Since there is no direct link between these two dates (except of course that they had to meet before they were married), your next question also seems unrelated. Have the lady jot down the number of years she's been married, up to the anniversary she will celebrate in the present year.

She gives the pad to her husband, who totals the numbers. When he announces the total, subtract the present year from his total. The result must be the year he met his wife.

Note that the two questions that must be answered accurately are directed at the wife because she is almost always certain to remember exactly how many years she has been married. The husband could jot down *any* year. Whatever date he writes, that is the date you reveal.

Here is an example. Assume the couple was married in 1972 and that the present year is 1982. She writes down the year she was married, he writes down the year he met his wife, and she then writes down the number of years she will be married when she celebrates this year's wedding anniversary. The result is:

She writes:	1972
He writes:	1970
She writes:	10

He then adds these numbers and gets 3952. When he announces this total, subtract 1982 (the present year) from it, getting 1970, the year he met his wife. You can use a pad and pencil to make the calculation, though with a bit of practice you should be able to do the figuring in your head. There is plenty of time to make the mental calculation because you are supposed to be traveling back in time by means of telepathy.

14 MYSTERIOUS MATH

This relatively simple trick opens the door to an age-guessing trick that can be developed into a major mystery. People like to

1	3	5	7	9	11	13	15
17	19	21	23	25	27	29	31
33	35	37	39	41	43	45	47
49	51	53	55	57	59	61	63

2	3	6	7	10	11	14	15
18	19	22	23	26	27	30	31
34	35	38	39	42	43	46	47
50	51	54	55	58	59	62	63

4	5	6	7	12	13	14	15
20	21	22	23	28	29	30	31
36	37	38	39	44	45	46	47
52	53	54	55	60	61	62	63

8	9	10	11	12	13	14	15
24	25	26	27	28	29	30	31
40	41	42	43	44	45	46	47
56	57	58	59	60	61	62	63

16	17	18	19	20	21	22	23
24	25	26	27	28	29	30	31
48	49	50	51	52	53	54	55
56	57	58	59	60	61	62	63

32	33	34	35	36	37	38	39
40	41	42	43	44	45	46	47
48	49	50	51	52	53	54	55
56	57	58	59	60	61	62	63

Fig. 10

hear about themselves, and if you can tell someone the year, the month, and the day of his birth, he can't help but be impressed.

Using slips of paper or pieces of cardboard, make up the six cards shown in Figure 10. Hand them to a spectator and have him think of a number smaller than 64. Ask him to find the cards which have his number. He hands you these cards and you immediately announce the number he is thinking of.

METHOD: The secret is well concealed. When the cards are given to you, simply note the first number on each card, that is, the number at the upper-left corner. Mentally add these numbers and the total will be the thought-of number. If, for example, he hands you the first, third, and fifth cards, the thought-of number is 1 + 4 + 16, or 21.

You can disguise the method by secretly marking the back of each card. Thus the back of the first card would have a secret marking indicating that the first number on this card is 1. The second card would be marked for the number 2, the next for the number 4, the next for 8, and so on.

Now when you are handed the cards which bear the spectator's thought-of number, you need not look at the faces of the cards. Hold them face down and slide your fingertips across the underside, saying that you can read the numbers by sense of touch. Pretend to concentrate so that the trick doesn't look too easy. Then reveal the thought-of number in dramatic fashion.

The best presentation is to have the spectator think of his wife's or his girlfriend's age. Have him find the cards which contain her age. When he hands you the cards, you immediately announce her age.

"Mysterious Math" uses a system of binary counting to arrive at the correct number. As you will see in the next two tricks, the mathematics can be completely concealed in novel ways.

15 ASTROLOGY CARDS

Here the basic idea of "Mysterious Math" will be developed to determine the day and month of a person's birthday. Hand the spectator the first five cards. The last card (the one with the number 32 in the upper-left corner) is not needed in this version.

Ask him to pick out the cards which have the day of his birthday. Thus if he was born on the 23rd, he would pick out all the cards that have the number 23 on them. He hands you these cards.

Add together the first number on each card he hands you. The total is the day of his birth, but don't reveal this information yet. All you know at this point is, for instance, that he was born on the 23rd. You are about to find out the month of his birth, but in an offbeat way. You do this by asking him for his astrological sign, that is, the sign he was born under. Most people know their sign, but for those who don't, show them the following table:

> March 21 through April 20: *Aries*
> April 21 through May 21: *Taurus*
> May 22 through June 21: *Gemini*
> June 22 through July 22: *Cancer*
> July 23 through August 23: *Leo*
> August 24 through September 23: *Virgo*
> September 24 through October 23: *Libra*
> October 24 through November 22: *Scorpio*
> November 23 through December 21: *Sagittarius*
> December 22 through January 20: *Capricorn*
> January 21 through February 19: *Aquarius*
> February 20 through March 20: *Pisces*

The intriguing point is that in most cases if you know the day of birth and the person's sign, you know what month he was born in! In our example you know from the cards that he was born on the 23rd. If he says his sign is Aries, he must have been born on March 23rd. He could not have been born on April 23rd because then he would be a Taurus.

There are four astrological signs that produce ambiguous results on rare occasions. If he was born on the 21st and says he is a Taurus, you don't know if he was born on April 21 or May 21, but a question will clear up the problem. Simply say, "You weren't born in April, were you?" If he says he was, announce that he was born on April 21. If he says he wasn't, you reply, "I thought so. My mental impression is that you were in fact born on May 21st." The ambiguous result occurs on only four days of the year in the signs of Taurus, Cancer, Leo, and Pisces, so chances are excellent that you will be able to reveal the spectator's date of birth exactly.

To review the trick, show the spectator the first five cards. Ask him to hand you the ones that contain the day he was born on. Then ask him for his astrological sign. Pretend to concentrate, then reveal the day and month of his birth. The table of all signs can be printed on the back of each card. This gives you an excuse to look at the back of the cards when they are handed to you.

If you like you can hold back this information and produce an even more astounding effect. Once you get the information, hand him all six cards and perform "Mysterious Math." Now you know his exact age as well. Thus, after the spectator looks at the cards twice, you can announce the day and month of his birth and how old he is. By subtracting his age from the present year, you know the year of his birth. Considering that the spectator merely looked at six cards twice, you have gathered a tremendous amount of information. An alternative approach is to perform "Astrology Cards" to learn the day and month, then perform "Guess Your Age?" to find out the spectator's age. When you have all this information you can reveal the month, the day, and the year of his birth.

16 NEALE'S NUMBERS

Robert Neale has taken the above idea and transformed it into a fiendishly clever magical routine. The same binary math underlies Neale's trick, but it is so ingeniously concealed that the trick would appear to have nothing whatever to do with mathematics.

Four word cards are used. The spectator mentally selects a word on one card and notes if it appears on the other cards. The magic now occurs in dramatic fashion. First one of the four cards vanishes completely. Then the writing on another card vanishes. Finally, without asking a question, the magician reveals the word merely thought of by the spectator.

METHOD: In fact only three cards are used. They are shown in Figure 11. Copy the words down on pieces of cardboard, making sure they are done in block letters so they can be read right side up or upside down. Cardboard is better than paper because it is easier to handle. Note in Figure 11 that some words are written in red and some in black.

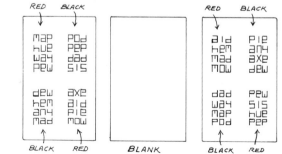

Fig. 11

Once the cards have been made up, stack them as shown in Figure 12. Then turn the packet over side for side so the writing side is down. Keep the packet squared so that the audience is unaware that you have only three cards.

You're now going to show the cards to the spectator. It will appear as if you show four different cards, but you will be showing the same two cards twice. Hold the packet in the left hand. Push the top card over to the right and flip it over side for side onto the packet as shown in Figure 13.

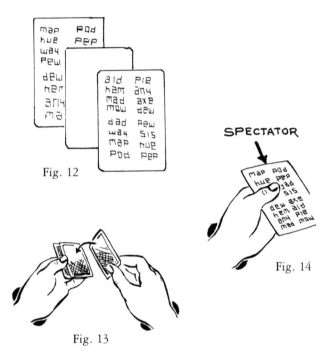

Fig. 12

Fig. 13

Fig. 14

Have the spectator mentally think of any word he sees on the card. Note that the spectator is seated across the table from you. His view of the card is indicated by the arrow in Figure 14. To see the card as the spectator sees it, turn this book end for end and view the card in Figure 14. Of course this is not necessary in the performance of the trick but it will convey the fact that the words appear different when viewed from different angles.

Have the spectator tell you the color of the chosen word. Remember the color. Turn the card over lengthwise (end for end) so it is writing-side down and place the card on the bottom of the packet. Say, "That's card number one." As you say this, transfer one card from the top to the bottom of the packet.

Turn over the next card side for side (sideways) as you did the first card. Have the spectator look over the words on this card. Tell him to find his mentally chosen word and to tell you the color of the word.

Once he has done this, turn the card down end for end or lengthwise. Place this card on the bottom of the packet. Say, "That's card number two." Transfer two cards from the top to the bottom of the packet.

Turn the next card over side for side. Ask the spectator to find his word and tell you its color. Turn the card over lengthwise so it is face down and place it on the bottom of the packet. Say, "That's card number three." Transfer three cards from the top to the bottom of the packet, one at a time.

Turn up the next card side for side. Have the spectator find his thought-of word and tell you its color. You're going to learn the thought-of word at this point. To do so we will have to interrupt the handling to explain how the key is determined. The key is:

First card:	Red-0	Black-1
Second card:	Red-0	Black-2
Third card:	Red-0	Black-4
Fourth card:	Red-0	Black-8

When you showed the spectator the first card, if he said his word was written in black, you secretly remembered 1. If he said black on the second card, you remembered 2. If he said black on the third card, you remembered 4. If he said black on the fourth card, you remembered 8. In each case the remembered number is added to any previous numbers. If he says the word is red at any time, add nothing to your total.

As an example, if the spectator said his word is black on the first and third cards, you would have remembered the numbers 1 and 4. Added together they total 5. If he saw his word on the second, third, and fourth cards, you would have remembered the numbers 2, 4, and 8. Added together they would have given you 14. This total then becomes your key.

We now return to the routine. You are holding the last card (supposedly the fourth) face up. When the spectator tells you the color of the word for the last time, the card facing you now has the word "mow" in the upper left corner. The subtle angle here is that *this* card tells you the thought-of word, as follows:

0 mow	1 pew
2 aid	3 hue
4 pie	5 way
6 axe	7 map
9 sis	8 mad
11 pep	10 hem
13 dad	12 any
15 pod	14 dew

For the key numbers 0 through 7, simply count from left to right, beginning at the top left word. For the key numbers 8 through 15, count from right to left beginning with the word "mad."

To take an example, the spectator might have said his word was written in black on the second and fourth cards. You would have remembered the numbers 2 and 8. The total is 10, so your key number is 10. Look at the word opposite 10 in the table above. The word is "hem." This is the spectator's word. The numbers are not on the card, of course, but they are easy to visualize when you mentally count to the chosen word.

You now know the spectator's word. Turn the card face down lengthwise and place it on the bottom of the packet. Say, "That's the fourth card." Transfer four cards one at a time from the top to the bottom of the packet.

Say to the spectator, "To find your word I'm going to send one card to a higher dimension." Deal the cards out in a row from left to right as shown in Figure 15. This is the first surprise. One of the four cards has completely vanished.

WORD	WORD	BLANK
CARD	CARD	CARD

Fig. 15

Say, "Now I'll keep one card here but send the writing to a higher dimension." Turn over the blank. It appears as if the writing has vanished.

"I'm getting a mental impression from the higher dimension." Pretend to concentrate. Turn the other two cards over, nod your head, then pocket the cards. Face the spectator and dramatically reveal the thought-of word.

17 STUNUMBERS

A classic problem in mind-reading tricks is to predict the total of a group of numbers chosen by the spectator. The following approach is undoubtedly one of the best methods in the literature. A combination of two subtle principles serves to conceal the mathematics completely. What the spectator sees is a series of honest steps which lead to an impossible conclusion.

Required for this excellent trick are four blank squares of cardboard or four sheets of paper. Ask the spectator to call out the ages of four people in his family. As he calls them out, write one age on each card.

Then write random ages of people in your own family on the opposite sides of the cards. The numbers are truly random in the sense that before the trick begins you have no idea which numbers you are going to write. In other words, you do not have set numbers in mind.

The spectator mixes the cards face up and face down. Your back is turned during the mixing of the numbers. You write something on a pad and hold it up with the writing side away from the spectator.

He totals the numbers he has chosen. They are random numbers and are not known to you at any time. He totals the numbers and announces the total. You turn the pad around. Written on it is the same total!

METHOD: The secret to this trick was developed by the Japanese magician Shigeo Futagawa. Whatever number the spectator chooses, write it on one side of the first square. When you turn the square over and write another number, simply write a number that is 15 larger than his number. If his number is, say, 8, then on the other side you write 23. If his next number is 12, the number you

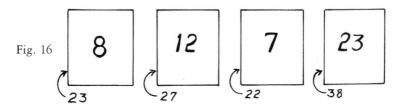

Fig. 16

write on the other side of this cardboard square is 27. A sample is shown in Figure 16.

After the cards have been filled out, ask the spectator to mix them. He can mix them any way he likes as long as he doesn't turn any cards over. When he has mixed them thoroughly, have him arrange them in a row on the table. Tell him to pick two cards and turn them over. Then have him pick two cards again and turn them over. They can be the same two cards or two different cards. Finally have him pick two cards and turn them over. Your back is turned and you have no idea which cards have been turned over.

You're now going to do a quick bit of figuring. When you were writing the spectator's numbers, you mentally added them together. Since the numbers will usually be small, they can be easily added together without pencil and paper. If you have trouble adding them, simply keep them in mind. When you pick up the pad to write your prediction, jot down the four numbers on the pad and add them up. This should not be necessary, but the option is always there.

When you've mentally totalled the spectator's numbers, add 30 to the result and jot down that number as your prediction. That's all there is to it. The rest is presentation. Just make sure that there is sufficient build-up before showing your prediction.

After the spectator has turned over cards two at a time as described above, have him total the numbers he sees. No one could say ahead of time what numbers will show at this point, but there will always be two of the spectator's numbers and two of yours, and further, the total will always match your prediction.

All that remains is to have the spectator announce the total. In our example the total will be 81. Turn the prediction around and show that it matches the spectator's total.

This trick is so strong you are likely to be asked to repeat it. Never repeat the same trick. Instead, switch to a trick that looks similar but is based on a different handling. An excellent encore to "Stunumbers" is "Lightning Strikes Thrice."

Another approach is to have the spectator choose four numbers as described above, then have him turn over cards two at a time (he does this three times). Then tell him to hide one of the numbers by covering it. Since each number on each card represents someone's age, he is in effect singling out one person. After he has done this, ask him to total the numbers showing on the other three cards and tell you the total. Then ask him to tell you which person is represented on the hidden card. He might say, "My nephew." At this you respond, "Oh yes, and he's 8 years old."

The secret is to mentally total the spectator's four numbers as explained above, and add 30 to the result. Jot down the sum. Whatever total the spectator mentions, subtract it from this number. The result is the hidden number.

For example, after he choses 8, 12, 7, and 23, and you add the appropriate numbers on the backs of the cards, he turns over cards two at a time, and does this three times. Then he hides the card with the number 8 on it. The numbers now showing on the other three cards might be 7, 27, and 38. He totals these and gets 72. In the meantime you've totaled his four original numbers and added 30 to get 80. When he announces a sum of 72, subtract this from 80 to get 8.

Don't reveal the number. Ask him which person in his family corresponds to the hidden card. He might say, "My nephew." You then respond, "Right, and he's 8 years old." By presenting the trick this way you are not revealing a mere number, but are doing the far more difficult trick of revealing the age of someone you've never met.

LIGHTNING CALCULATORS

The true lightning calculator has the ability to perform prodigious feats of arithmetic without recourse to pencil and paper. Almost all lightning calculators resort to tricks and shortcuts on occasion to produce seemingly incredible results. Most of these methods are well known and have been recorded in the literature on the subject.

The present chapter considers some of the less well-known methods of producing instantaneous answers to complex arithmetic problems. They can be performed as evidence of exceptional mental prowess or they can in some cases be performed as tricks involving supernatural powers. The chapter closes with a discussion of a system of making any large number evenly divisible by numbers between 1 and 10. The subject is well known but this is believed to be the simplest system currently available.

18 LINEAR A

This little-known trick of the lightning calculator appears to produce an impossible result because you are able to make a giant 18-digit number exactly divisible by 7 *at a glance.* Have the spectator jot down the digits in his Social Security number in any order. Then have him repeat this series of digits alongside the first number. If his Social Security number is 123-45-6789, he might scramble the digits and write the new number twice, as follows:

867594312867594312

Tell him that you will erase two of the digits in this giant number, replace them with two new digits, and the new number will be evenly divisible by 7. When you get the sheet of paper from the spectator, erase the second and eleventh digits, replace each of them with a 7, and incredibly enough, *that's it.* The new 18-digit number will be evenly divisible by 7. The spectator will not believe that you could figure the quotient that quickly. From his point of view it is impossible. But when he performs the division he should be astonished to discover that the giant number is now divisible by 7.

An interesting variation is this. After he has jotted down the giant number, ask him for his lucky number. Whatever the digit, erase the second and eleventh digits and replace them with the digit called out by the spectator. Tell him that if this is indeed his lucky number, the giant number will be evenly divisible by 7. When he performs the division, he'll find that the number is in fact evenly divisible by 7.

The trick will work with any nine-digit number, but by using the spectator's Social Security number and tying it in with his lucky number, you direct attention to the spectator. He must conclude that it works only with his number. Just make sure that whatever nine-digit number he writes, he repeats it exactly. Neal Thomas points out that if there are 0's in the number, the spectator should change them to some other digit. This gets around the problem of a nine-digit number ending in 0, because the trick will not always work with such a number.

Since the feat appears impossible you can exploit this point by setting up the trick as a game. Remark that when the Minoan civilization was uncovered early in this century, it was found that they invented a superfast method of dividing numbers called Linear A. Have the spectator jot down his Social Security number in scrambled order, then have him repeat this scrambled number. The result will be an 18-digit number. When you get the paper, change the second and eleventh digits to 7 as described above. Turn the paper over and say, "I'm positive that this number is now divisible evenly by 7. I'm so positive I'm willing to bet on it."

Since it appears impossible, the spectator is likely to bet that you're wrong. He checks the math and finds that he has lost the bet. Asked to repeat the stunt, you may want to switch to "Slippery Sevens," described next, because although it involves division by 7, it is not related to "Linear A."

19 SLIPPERY SEVENS

If you have access to a calculator or office machine with a large readout capability, you can perform a strange and baffling number feat. Explain to the spectator that you can quickly calculate a number that will either be exactly divisible by 7 or will be off by no more than 1. Thus the number 63 is evenly divisible by 7 but 55 is one less, and 36 is one more.

The numbers you provide may be as small as 3484 or as large as 111,284,641 or even larger, depending solely on the capacity of the calculating machine. Further, there is no apparent pattern to the numbers. Making just a quick calculation on the machine, you can produce numbers of gargantuan size, all of which are within 1 of being evenly divisible by 7.

The odd angle on all of this is that even you don't know if the number is going to turn out to be exactly divisible by 7. The trick here is different from "Linear A" on precisely this point: the number you produce *may* be divisible by 7, but all you can say with certainty is that it will never be more than 1 off.

METHOD: The first time I came across this startling secret was in a book by Professor Hoffmann. It seems to be very little known. Richard Guy provided a proof that it works in all cases.

The idea is to enter any number into the calculator, multiply it by itself, then multiply it by itself again. In other words you cube the number or raise it to the third power. The result will be a number that will be within 1 of being evenly divisible by 7.

For example, if you choose 29, the product $29 \times 29 \times 29$ will be 24,389, a number that is one greater than a multiple of 7. If you choose 481, the product $481 \times 481 \times 481$ is 111,284,641, a number greater than the capacity of most pocket calculators but well within the range of office machines. This number is one less than a multiple of 7.

If you have access to a machine that can be programmed, arrange for it to cube any number entered into it. Then have the spectator enter any random number. The machine cubes the number (without the spectator knowing precisely how the number is produced) and the spectator divides this new number by 7. Here the trick would be presented as one where you taught a machine to convert any number into one that will be within 1 of being evenly divisible by 7. In a sense, this is true.

20 ALBERTI'S GAME

The clever shortcut concealed in this method allows you to perform a stupendous feat of mental arithmetic with virtually no effort. Ask someone for a three-digit number. Say he names 593. Write it twice:

593   593

Ask him to name another three-digit number. Say he names 482. Write it under the number at the left:

593   593
482

Finally you add another three-digit number, writing it under the number at the right, as follows:

593   593
×482   ×517

Now you demonstrate that you can mentally perform both multiplications and add the two products together in half the time it takes the spectator, *even if he uses a calculator.* Indeed, the sum of the two products can be found as quickly as you can write them down.

METHOD: When the spectator gives you the number 482, find the 9 complement of this number and write it as the second multiplier. To find the complement, subtract each digit from 9. If the spectator's number is 482, the 9 complement is 517. This number is jotted down as the second multiplier. If you work with a confederate, have him give you this multiplier. Then it appears as if all four numbers are random.

While the spectator is busy multiplying 593 by 482, then 593 by 517, and then adding the two products, you resort to the following shortcut. Subtract 1 from 593 and jot down this number, 592.

Then find the 9 complement of this number by subtracting each digit from 9. The result in this example is 407. Write down this number to the right of 592, getting 592,407.

Surprisingly enough, this number is the sum of the two products.

The trick works with four-digit numbers, five-digit numbers, and so on, but the spectator confronted with large numbers is likely to make a mistake. Using three-digit numbers the spectator can obtain the correct answer quickly and the problem still looks impressive.

21   THE FIFTH POWER

The spectator chooses any number from 1 to 100 and multiplies the number times itself four times. He announces the result and you immediately name the original number.

1	100 THOUSAND
2	3 MILLION
3	24 MILLION
4	100 MILLION
5	300 MILLION
6	777 MILLION
7	1 BILLION, 500 MILLION
8	3 BILLION
9	6 BILLION
10	10 BILLION

Fig. 17

METHOD: It is necessary to memorize the simple table shown in Figure 17. It is easy to find patterns that aid the process. For example, beginning with the first number, every third number begins with a 1. Every number after that begins with a 3, and so on. The table allows for a rapid computation of the given number as follows.

Let's say the spectator chose 57. He would multiply 57 × 57 × 57 × 57 × 57. He then announces that the grand product of his multiplication is 601,692,057. When he says, "Six hundred and one million," you know immediately that the number lies between 300 million and 777 million in the table. Always pick the smaller number. In this case the smaller number is 300 million and the digit associated with it is 5. Thus you know that the first digit of his original number is 5.

Wait until he announces the final digit of the grand total. That digit is the second digit of his original number. In our example the final digit of the product is 7, so you know that the original two-digit number chosen by the spectator is 57.

There are systems for finding square roots and cube roots but they require the memorization of specific numbers. In the above system the subtle angle is that while it appears as if it is tremendously difficult to find the fifth root of a number, it is actually *easier* to find fifth roots than square or cube roots. Note too that you listen only to the first number he calls out and the last digit. He can even lie on intervening digits, though this tends to give the secret away. However you present it, the demonstration is bound to impress the audience.

If the spectator's answer is less than 100,000, the first digit of the spectator's number is 0. For example, if he calls out 16,807, the chosen number is 07 or just 7.

22 PLAYING THE MARKET

Stock investors used to resort to a doubling system to cheat the market, but nowadays they have more subtle ideas which link two or more stocks together. To illustrate the system, have the spectator jot down the names of two prominent stocks. He might choose General Motors and A.T.&T. He lists them at some opening price, say $46 and $4, respectively. They are entered on a pad as shown in Figure 18.

To figure out what the combined price will be on Monday, the spectator adds together the two stock prices. In this case he gets 46 + 4 = 50 and enters this on the line marked Monday.

Then he adds the two current numbers and enters this on the line marked Tuesday. In this example he would add 4 + 50 and get 54. He continues in this way, getting a new stock price for each day. After he reaches the final price, ask him to add together all the numbers on the pad. He does, and he is surprised to discover that you've written exactly the same total long before he did.

G.M.	46
A.T.&T.	4
M.	
Tu.	
W.	
Th.	
Fr.	
Sa.	
Su.	
M.	

Fig. 18

G.M.	46
A.T.&T.	4
M.	50
Tu.	54
W.	104
Th.	158
Fr.	262 ⬅
Sa.	420
Su.	682
M.	1102

Fig. 19

METHOD: Mark off spaces on the pad as shown in Figure 18. Have the spectator enter two stock prices. They can be any numbers but to make the later work easy, have him use relatively low numbers at the start. He adds these two numbers together. Whatever the result, he enters it on the third line. Then he adds the numbers on the second and third lines (4 and 50 in our example) and enters this number on the fourth line. He adds together the numbers on the third and fourth lines (50 and 54 in the example) and enters the total on the fifth line. The process continues until he has filled in the final line. A sample is shown in Figure 19.

Hand him a pocket calculator. As you do, glimpse the entry next to Friday. In the example of Figure 19 this number is 262. Tell the spectator to add up the ten entries with the aid of the calculator and announce the total.

While he does this, pick up a piece of paper, jot down the Friday entry and multiply it by 11. The procedure is always the same, and because multiplication by 11 is so easy you can do it instantly. In the above example the figuring looks like this:

$$\begin{array}{r} 262 \\ \underline{262} \\ 2882 \end{array}$$

When the spectator adds up the numbers, he will arrive at a total of 2882 also. But he arrives at the total long after you do, and you don't use a calculator. Simply explain that you learned the stock-market system years ago and were sworn to secrecy.

One tip on presentation. When the spectator adds up the numbers, tell him to start at the final entry and work up to the first entry. If he begins at the first entry and works down, he may become aware of a slight pattern to the way the numbers increase, but from the bottom up there is no apparent pattern.

23 TRICKY TOTAL

The magician has several spectators jot down three-digit numbers on slips of paper. The slips are gathered in a hat, mixed, and two slips removed. A spectator with a large pad writes the first number, which might be 682. Under it he writes the second number, which may be 143. Then he multiplies the two numbers.

No matter how quick the spectator is, he can't beat the magician, who makes a brief notation on a scrap of paper and instantly announces the product of the two numbers.

METHOD: In the hat, tucked under the hatband, is a slip of paper bearing the number 143. That is the only preparation.

Hand three or four slips of paper to each of several spectators. Have them jot down three-digit numbers, one three-digit number on each slip. Gather the papers in a hat. Have a spectator mix the papers. Then reach into the hat and remove the slip that was hidden under the hatband.

Give this slip of paper to a spectator. Have him jot down the number on a large pad. Then reach into the hat and withdraw any other slip. Note the number on this slip as you hand it to the spectator. Tell him to jot down this number under the first number and multiply the two numbers.

Pick up a blank piece of paper and perform the following shortcut method of multiplying the two numbers. Jot down the second number twice. If the second number was 682, write 682,682. Quickly divide this number by 7. Whatever the result, announce it. In this case the result will be 97,526.

When the spectator finishes multiplying 143 by 682, he will discover that the total is 97,526, exactly as you foretold.

Note that when you divide the number by 7 the result will always come out even. If you get a remainder, you have made a mistake. Even if you have to go back and divide the number by 7 correctly, you should come up with the right answer long before the spectator completes the problem.

The reason for handing him a large pad is that you want him to make the numbers large and clear so they are easily visible to others. They can thus check on his multiplication as he goes along. Naturally if he makes a mistake it will take him that much longer to find the correct product. This makes the problem seem that much more difficult. By contrast, your ability to announce the product almost instantly appears to border on the impossible.

24 MATHEMATICALLY INCLINED

This mathematical oddity lies midway between a trick and a puzzle. Begin by asking a spectator to jot down a row of 10 digits. Let's say he chooses 5 7 4 3 1 8 2 9 7 8. Glancing at the row of numbers you write a prediction on a piece of paper, fold it and place it to one side.

The spectator is asked to begin at the left side of the row, take each pair of adjacent digits, add them together, and subtract 9 from the result. The first two digits are 5 and 7. Adding them together gives you 12. Subtracting 9, you get 3. Write the number 3 below the original row of digits, between the 5 and the 7.

The next set of adjacent digits is the 7 and the 4. Add them together, getting 11, then subtract 9 to get 2. The 2 is written below the original row, between the 7 and the 4. Continue until you've covered every pair of digits in the row. The result will look like this:

5 7 4 3 1 8 2 9 7 8
3 2 7 4 9 1 2 7 6

The spectator now performs the same mathematical operation on the second row of digits and continues the process with the third row, the fourth, and so on. The final array will look like Figure 20. Note that, for totals less than 9, you don't subtract nine.

The final digit is a 1, arrived at only after several minutes of effort on the spectator's part. Yet at the beginning of the trick you jotted down a number on a slip of paper, and when that slip is opened it is seen that you correctly predicted that the final digit would be a 1.

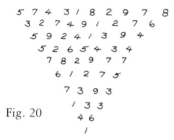

Fig. 20

METHOD: This routine was suggested by L. Vosburgh Lyons for an array known as Pascal's Triangle. Your brief glance at the original row of numbers is directed at certain digits, specifically the first, fourth, seventh, and tenth digits. Mentally add the fourth and seventh digits together and multiply the result by 3. To this add the first and tenth digits.

In the example you would add 3 + 2, getting 5, and multiply this by 3 to get 15. To this you add 5 + 8, getting 28. Now subtract from this the highest multiple of 9 that is smaller than 28. Here the highest multiple is 27. Subtracting 27 from 28 gives you a remainder of 1. This is the number you jot down on the paper.

To disguise the figuring, previously write on the paper, "You will arrive at the digit —." Leave a blank space. Do the figuring as you pretend to write the prediction, then fill in the required digit in the blank space, fold the prediction, and place it aside.

There is just one more point. If the spectator adds two digits together and the result is less than 9, he does not subtract 9 from it. Instead he simply writes the number as is.

The trick can also be done with playing cards. Ten cards are chosen at random (10's and picture cards are excluded), shuffled, and dealt out in a row. From this a final digit is arrived at as described above. If playing cards are going to be used, the more subtle approach is to have the deck shuffled and spread face down on the table. You and the spectator take turns drawing cards from the deck until a total of ten cards have been withdrawn. These cards are arranged in a row and the spectator then does the calculation described above to arrive at a final digit. In this case the prediction is seen to be taped to the ceiling from the beginning. When it is opened and read, it correctly predicts the number the spectator would arrive at.

The ploy is to mark the backs of four known cards, say any Ace, Five, Eight, and Four. Pick these four cards out of the deck after it has been shuffled and spread face down. When you deal the ten cards out in a row, make sure the marked cards go as follows: Ace in the first position, Five in the fourth position, Eight in the seventh position, and Four in the tenth position. Now the spectator must arrive at a result of 8, which of course matches your prediction.

25 FAKE ARITHMETIC

This is a subtle method for adding five five-digit numbers instantly. Although based on established principles, the application is offbeat and well concealed. The subtle point is that the spectator does all of the hard work for you.

Several spectators call out five-digit numbers. You list them one under the other until you have five such numbers. Explain that you are going to add them up instantly, but to emphasize how difficult the feat is, you have the spectator try it with different numbers. You jot down five five-digit numbers and ask him to try his hand at it. Eventually he will struggle through to the total. Then you return to the original problem and *instantly* produce the correct sum.

METHOD: Ask for five five-digit numbers and jot them down as they are called out. An example is given in Figure 21. Remark that to underline the difficulty of performing lightning calculation you will give the spectator an addition problem to compare how long it takes him to solve a similar problem.

ORIGINAL
PROBLEM

2 1 3 9 8

4 1 7 6 2

5 8 3 2 9

8 1 6 3 4

7 2 9 5 6

Fig. 21

SPECTATOR
SOLVES THIS
PROBLEM

7 8 6 0 1

5 8 2 3 7

4 1 6 7 0

1 8 3 6 5

2 7 0 4 3

4

Fig. 22

Here is where you cheat. The five numbers you jot down are the 9 complement of the five numbers given to you in Figure 21. This means that each digit of the second problem is obtained by subtracting the corresponding digit of the original problem from 9.

In our example the first number given to you by the spectator is 21,398. Subtract the 2 from 9 to get 7. Subtract the 1 from 9 to get 8. Subtract the 3 from 9 to get 6, and so on. The result will be the number 78,601. This is the first number you write for the spectator's problem. Continue with each of the other numbers, jotting down their 9 complement to form new numbers. To all of this add 4. The result will be the problem of Figure 22.

The audience assumes you are merely jotting down another addition problem. With practice you will be able to do it quickly, forming the new numbers merely by glancing at the original problem.

Have the spectator add up this column of numbers. He will get 223,920. Make sure he adds the numbers properly. Occasionally glance at your watch to emphasize how long it is taking him to add the numbers.

To find the sum of the original numbers, subtract each of the last five digits on the right in the spectator's answer from 9. The first digit, that is, the digit at the far left, is subtracted from 4. The result is the total of the numbers shown in Figure 21. In our example the spectator's total is 223,920. Beginning at the far right, subtract each of the first five digits from 9 and write down the result. Subtract the leftmost digit (the 2) from 4 and write down the result. The outcome is 276,079, which is the correct sum of the numbers in Figure 21.

26 ULTRAFFINITY

This is a method of allowing you to divide any large number evenly by any number from 1 to 10, except 7. There are well-known systems for accomplishing this but the system described here is believed to be the easiest in print. No claim is made for mathematical elegance. The idea is to solve the problem rapidly and accurately by the simplest route; that is the system that will be described here.

A sample effect is the following. Tell the spectator to remove a dollar bill from his pocket and jot down the serial number on a slip of paper. He then removes another dollar bill from his pocket, and jots down that number to the right of the original number. The result will be a giant 16-digit number.

He then calls out any number between 1 and 10, except 7. You tell him that as soon as you hear this number, you will glance at the 16-digit number and add a number to it such that the new number will be evenly divisible by his number.

Assume he calls out 6. You immediately add a number to the 16-digit number. When he divides this large number by 6, he will find that it is exactly divisible by 6.

METHOD: There are several secrets involved here. The basic idea is to make the work as easy as possible by combining different schemes. We will not consider divisibility by 7 because all such procedures are too complex. If the spectator names 7, tell him you'll postpone that for the moment. Have him call out another digit. Then you get back to 7, but perform "Linear A" (No. 18) or "Dollars and Dice" (No. 27). For any other digits between 1 and 10 the method is as follows.

If the 16-digit number is to be evenly divisible by 3, 6, or 9, add together all 16 digits and reduce this result to a single digit. Say the spectator wrote this 16-digit number:

$$9,834,271,207,621,388$$

Adding together the 16 digits, you would arrive at a total of 71. Adding these two digits, you arrive at 8.

If the result you get is even, subtract it from 18 and have the spectator append that number to the right side of the given 16-digit number. If the result you get is odd, subtract it from 9 and have the spectator append that number to the right side of the given 16-digit number.

In our example the result of adding the digits is 8, an even number. Subtracting it from 18, we arrive at 10. Append 10 to the right side of the given 16-digit number. The result will be:

$$983,427,120,762,138,810$$

This giant 18-digit number will be found to be evenly divisible by 3, 6, or 9.

If the spectator wants the given number to be evenly divisible by any other digit between 1 and 10, have him append 6240 to the right side. When he has done this, the new number will be evenly divisible by 1, 2, 4, 5, 8, or 10.

It can be seen that you need remember only two rules. If the number is to be made evenly divisible by 3, 6, or 9, sum the digits and reduce this to a single digit. If the single digit is even, subtract it from 18 and add the result to the right side of the given 16-digit number. If the single digit is odd, subtract it from 9 and add that to the right side of the given 16-digit number. For any other divisor, simply add the number 6240 to the right side of the 16-digit number.

There is an interesting variation in which you make the given number divisible by *all* digits between 1 and 10 (except 7). When the spectator jots down the given 16-digit number, mentally sum the digits. Determine the number to be added to the right side by using the rule for divisibility by 3, 6, or 9. Add that number to the right side. Then add 720 to that. The resulting number will be evenly divisible by *any* digit (except 7) between 1 and 10 or by the product of any two digits.

This method can also be used as a prediction. Mentally sum the digits in the 16-digit number and reduce the sum to a single digit. Then jot down a number as determined above which will convert the 16-digit number into one that is evenly divisible by any digit from 1 to 10. Append this number to the right side of the 16-digit number. Only then do you have the spectator call out a digit from 1 to 10. That digit will of course divide evenly into the new number.

This explains how to solve the problem that we started with, that of making a 16-digit number evenly divisible by any digit between 1 and 10. We did not cover the case of divisibility by 7. There is no simple method in this case, but you can avoid the problem completely with the following routine.

27 DOLLARS AND DICE

In this impressive trick you tell a spectator to remove a dollar bill from his pocket and jot down the serial number on a slip of paper. Then have him roll a die. He can use just the top number of the die or the top number added to a number on an adjacent side. Say he rolls a 6 and decides to use just that number.

You glance at the serial number, then instantly add a number to it such that the new number will be exactly divisible by 6.

As a topper, you tell him to write his serial number in reverse order. Without looking at the new number, you add a number to it such that the new number is still evenly divisible by 6. And remember that you do it without knowing the reversed serial number!

METHOD: The secret is a combination of "Ultraffinity" and a swindle that makes a simple trick seem like an incomprehensible mystery.

When the spectator jots down the number on the dollar bill, mentally total the digits as he writes them. Remember the total. Then turn aside, remove the die from your pocket and hand it to the spectator. Walk away and ask the spectator to roll the die on the table.

Ask him to give the die one final roll. Explain that he can use just the top number or, if he feels that is too easy, he can use the top number plus the number on any adjacent side of the die. Whatever he decides, tell him you will instantly add digits to his serial number such that the new number will be exactly divisible by the number he rolled on the die.

Of course the spectator will not settle for an easy divisor like 2 or 5, but you can see that it really makes no difference because the hard work has already been done. Whatever divisor he names, follow the procedure of "Ultraffinity" to establish the number that must be appended to the right side of the serial number. Since you've already added the digits mentally, the rest of the procedure is virtually instantaneous. You have merely to try this one to convince yourself of the startling impact it has on laymen.

"Ultraffinity" gave no procedure for divisibility by 7. The reason is that the procedures are without exception complex. You don't want it to appear to be a computational problem, so if you ignore the possibility of divisibility by 7 you have only to glance at the serial number to cover all other possibilities.

You might wonder what would happen if he added together two numbers on the die and came up with 7, but that is why a die is used; with any standard die it is impossible to sum two adjacent numbers and arrive at 7. The spectator never questions the point. As far as he is concerned, he has a perfectly free choice of any number between 1 and 12.

The other problem is divisibility by 11. It is possible that the spectator will roll a 6 and decide that he wants to add the adjacent 5 to it. Alternately, he may roll a 5 and decide that he wants to add the adjacent 6 to it. Either case will result in the problem of making the serial number evenly divisible by 11. It can be done, but you must go back to the dollar bill and perform an add-subtract operation. As with the problem of divisibility by 7, the method is complicated and not worth the trouble.

If the spectator rolls a 5 or a 6, have him roll the die again. If he rolls again and gets a 5 or a 6 the second time, chances are small that he will add a 6 to the 5 or a 5 to the 6 to end up with the troublesome divisor 11. Still, he may insist that he wants the divisor to be 11. In this case simply bluff and call out any number. You may be lucky. But even if you get it wrong, the spectator will be impressed by the attempt. Ask for another try. When he rolls the die this time to get a new divisor, you must succeed.

The final point is the kicker at the finish of the trick where he scrambles the digits and you still make the number evenly divisible by the number he rolled with the die.

If you are required to make the scrambled number divisible by 3 or 6, have him append a 6 to the right side. For divisibility by 4 or 8, have him append 624. For divisibility by 2, 5, 9, or 10, have him append 90 to the right side.

It seems as if this problem is impossible because you are trying to alter a number without knowing what that number is, but in fact all of the difficult work has been done in the first part of the routine.

If you are willing to take a chance, there is a form of the repeat that works almost all of the time. You can even handle it as a prediction since the information can be written out in advance. After the spectator has scrambled the digits, have him add 720 to the right side *regardless* of the divisor. The new number will then be evenly divisible by whatever number he rolled on the die.

MAGIC WITH NUMBERS

This is the first chapter in the book to include some tricks where cards are the major item of apparatus. While there are many fine mathematical card tricks, the card effects here are limited to those routines which depend on number magic to work. The noncard routines are outstanding examples of how numerical methods can be logically woven into mysterious magical effects.

28 MENTAL AMAZEMENT

This routine is in part a number trick and in part a card trick. Based on an idea of Harry Canar, the mathematics is well hidden by the introduction of a second selected card. As the audience sees it, two cards are chosen by a random process. The magician then proceeds to reveal each card.

To describe the effect is also to describe the method, so we shall proceed immediately to the way the trick is done. Before starting, remove any easily remembered card, such as the Two of Spades, and place in in your pocket. This is the only preparation. If you use borrowed cards, the Two of Spades can be secretly placed in the pocket when you are in a room away from other guests.

To present the trick, hand the deck to anyone and have him shuffle it thoroughly. Ask for a number from 1 to 25. Whatever number is called out, tell the spectator to count that many cards off the top of the deck, note the card he counted to, and place that card in his pocket.

Have him jot down the value of his card on a piece of paper (Jacks = 11, Queens = 12, Kings = 13). To this he adds the number that was called out. If the number was 14 and he pocketed the Three of Diamonds, he would add 14 + 3 to arrive at a total of 17.

Tell him to multiply the total by 10. In the example he would multiply 17 × 10 to arrive at 170.

Have him add 1 to the total if he chose a Club, 2 if he chose a Heart, 3 if he picked a Spade, and 4 if he picked a Diamond. In our example he picked a Diamond, so he adds 4 to 170 to arrive at 174.

Now someone in the audience calls out any number between 25 and 52. Have the spectator count to that number, remove the card he counted to and hand it to you sight unseen. Without looking at the card you place it in your pocket.

Whatever the second number was, he adds it to his previous total. If, for example, the second number was 48, he adds 174 + 48 to arrive at 222. Have him tell you this total.

Unknown to the spectator, you have been doing some simple mental calculations as the trick proceeded. When you heard the first number, 14, you multiplied it by 10 to arrive at 140. When you heard the second number you added it to 140. In our example you heard 48 as the second number, added it to 140, and obtained 188. This is easy to do without pencil and paper.

When the spectator calls out the grand total, merely subtract your number from his number. In our example the grand total is 222, so you obtain 222 − 188 = 34. But *this* tells you the identity of the card in the spectator's pocket! The digit on the left tells you the value of the spectator's card, in this case 3. The digit on the right tells you the suit. Since this digit is 4 and because 4 = Diamonds, you know the card was the Three of Diamonds.

Don't reveal the spectator's card yet. Pretend to concentrate, then say, "I think you gave me the Two of Spades." Remove the Two of Spades from your pocket and show it. Then say, "And I think you gave yourself the Three of Diamonds." The spectator should be flabbergasted when you correctly reveal the names of both cards.

29 LOCKED-ROOM MYSTERY

The magician reenacts a crime that was committed in a hotel. To do so he draws a 3 × 3 layout of the rooms and then numbers the rooms as shown in Figure 23. The crime was the theft of an invalu-

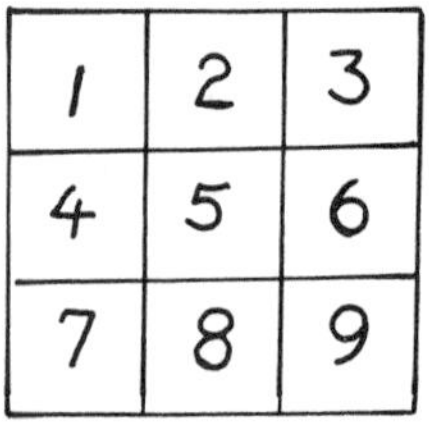

Fig. 23

able ring, so the spectator is asked to remove his finger ring and place it on any square.

The spectator is then given a card with instructions that explain how the police used clever deduction to find out which room contained the ring.

The instructions on the card read:

1. Cross out the 6.
2. Move seven times and cross out the 7.
3. Move four times and cross out the 3.
4. Move six times and cross out the 1.
5. Move five times and cross out the 8.
6. Move two times and cross out the 2.
7. Move once and cross out the 9.
8. Finally, move seven times and cross out the 4.

A move consists of transferring the ring to the adjacent square above, below, left, or right. He can't move diagonally, and the ring can't come to rest on a crossed-out square.

After the spectator has made the above moves, shuttling the ring from square to square, it may finally come to rest on Square No. 5. Turning over the card with the instructions, it is revealed that there is a message which reads, "The police found the ring in Room No. 5."

METHOD: This ingenious routine was invented by Martin Gardner. The secret depends on the fact that, unkown to the spectator, there are really two sets of instructions. If the spectator places his ring on an odd-numbered square, bring out the instructions listed above.

If he places his ring on an even-numbered square, bring out the following set of instructions:

1. Cross out the 7.
2. Move four times and cross out the 3.
3. Move seven times and cross out the 2.
4. Move three times and cross out the 9.
5. Move one time and cross out the 8.
6. Move two times and cross out the 6.
7. Move five times and cross out the 1.
8. Move three times and cross out the 4.

The prediction on the back of each instruction card reads, "The police found the ring in Room No. 5."

Draw the layout and have the spectator place his ring on any square. If he places it on an odd-numbered square, bring out the first set of instructions and place them before the spectator without letting him see the message on the back. If he places the ring on an even-numbered square, remove the second set of instructions from your pocket and place them before the spectator.

Have the spectator follow the instructions on the card. He can move the ring randomly from square to square but at the end of each move the ring must come to rest on a square that has not been crossed out. After the final move is completed, the ring will invariably rest on Square No. 5.

If you want to go to some slight extra trouble, there is a startling variation that may be of interest. Instead of using the spectator's ring you use a key that has a plastic tag attached to it. The tag is the kind that can be found on hotel-room keys. It has the number 5 printed on it. Unhook the tag, taking care not to show the number printed on it. Then have the spectator place the key on any square. Explain that the police found the key in one of the rooms and they knew that the key came from the room where the crime was committed. They suspected the key could lead them back to that room.

When the spectator places the key on a square, remove the proper set of instructions from the pocket. There is no message on the back of either instruction sheet. The spectator moves the key from room to room, finally ending up in Room No. 5. Say, "The key did lead them back. Look—." Here you turn over the plastic tag to reveal the number 5.

30 THE TWO GUNS

If you are asked to repeat the above trick you might want to consider a splendid variation devised by Robert Neale. In Neale's strangely concocted version, the mathematical nature of the trick is neatly concealed.

Neale uses a small black toy gun to represent the bad guy and a silver gun to represent the good guy. If these props are not readily available you can use a penny and a dime.

The layout is shown in Figure 24. The good guy and the bad guy chase one another from state to state. They are destined to meet and the magician successfully predicts which state they will meet in.

MONTANA	UTAH	ARIZONA
CALIFORNIA	IDAHO	COLORADO
NEW MEXICO	NEVADA	WYOMING

Fig. 24

The state you predict is Idaho. At the start place the folded prediction slip under a cup. The prediction can be a map of Idaho or simply a scribbled note that the two gunslingers were destined to meet in Idaho.

Turn your back. Have the spectator place the black gun on any state and the silver gun on any state. Now he moves each gun according to the number of letters in the state it rests on. If the state, for example, contains five letters, he would make five moves. As before, each gun can be moved right or left, up or down, but not diagonally. After this move, he is to cross out Montana and Wyoming.

When this has been done, direct the spectator to move the black gun three times. Have him move the silver gun three times. Each gun must land on a state that has not been crossed out. When he has made these moves have him cross out Colorado and California.

Direct him to move each gun three times. Then have him cross out New Mexico and Arizona.

Have him move each gun three times. He is then to cross out Nevada and Utah.

Although the guns started out in random states and made random moves while the chase was on, both guns will now be in Idaho, exactly as foretold by the prediction.

The trick works automatically. Instead of crossing out each state as directed above, the spectator can cover each of the indicated states with a coin. You explain that each coin represents a reward posted by that state for the apprehension of the criminal.

Neale has suggested a number of different plot ideas for this trick. In one, a boy and a girl pass through many different emotions (written on the squares) until they meet in true harmony on the center square. The trick is easily made up, the apparatus can be carried in the wallet or pocket, and the routine itself is an impressive mystery.

31 PRIME CHOICE

Prime numbers have the characteristic that they are not divisible by any number (other than 1 and themselves). A number of intriguing tricks have been devised which depend on prime numbers for their working. One of the best is this trick.

The spectator chooses a card, places it on top of a packet, and cuts the packet. He turns up the top card. Whatever its value, he deals that many cards from top to bottom. Then he turns up the new top card. Whatever its value, he deals that many cards from top to bottom.

The process continues. It is seen that all of the cards turned up are indifferent cards. The spectator does in fact turn up all but one of the cards. That final card will always be the chosen card. Note that the entire trick takes place with the cards in the spectator's hands.

METHOD: The idea behind this routine was independently developed by Jacob Daley and George Sands. The trick works with a prime number of cards. In this case we use 13 cards. Begin with 12 cards arranged as follows from top to bottom: 10-5-7-2-3-8-Jack-9-6-Queen-Ace-4.

Ask a spectator to choose any card from the balance of the deck, note it and place it on top of the packet. Then have him cut the packet and complete the cut to lose his card. He may give the packet several straight cuts. At this point the packet is face down. No one knows the location of any card.

The spectator turns the top card of the packet face up and leaves it on top of the packet. Assume the card is an 8. He deals cards one at a time from the top to the bottom of the packet, beginning with the face-up card, until he has dealt a total of eight cards.

Then he turns the new top card face up. Whatever its value, he deals that many cards from top to bottom, beginning with the face-up card on top of the packet. In this example the new top card would be a 10. He deals ten cards from top to bottom, dealing one card at a time, beginning the count with the face-up 10-spot.

He continues in this way until there is just one card face down in the packet. This card will be the chosen card.

Remember that Jacks have a value of 11 and Queens a value of 12. On rare occasions the spectator will cut the packet the first time, complete the cut, turn up the top card, and it will be his card. Congratulate him on his amazing ability to cut to his own card. If he cuts to some other card, say to him, "I'll bet that your card is the last card you turn face up." You can't lose.

32   LUCKY 13

If asked to repeat "Prime Choice," you can do an even stronger version in which the packet is shuffled by the spectator. This variation, invented by George Sands, goes as follows.

The spectator chooses a card. It is placed in the packet. The spectator calls out any number between 1 and 12. Say he calls out 5. The performer counts down to the fifth card and turns it face up. It is not the chosen card.

Leaving this card face up, the magician counts down to the new fifth card and turns it face up. It isn't the chosen card. He continues in this way, counting to the fifth card each time. The chosen card doesn't turn up until the last card.

METHOD: At the start of the trick remove any twelve cards from the deck. If you have just completed "Prime Choice," twelve of these cards can be used. Let the spectator shuffle the packet.

Have the spectator choose a card from the balance of the deck. With the chosen card in one hand and the twelve-card packet in the other, place both behind the back, saying you're going to insert the card into the packet. In fact put it on top of the packet.

Remark that his card will be the last card you turn up. Have him call out a number from 1 to 12. If the spectator says 5, deal the top four cards one at a time from top to bottom. Turn up the fifth card and leave it face up on top. Ask the spectator, "Was this your card?" He will say no. Reply, "I thought so," and transfer this card, still face up, from the top to the bottom of the packet.

Now repeat the process. Deal four cards from top to bottom, turn up the fifth card and transfer it to the bottom of the packet. Continue until there is just one card left face down. It will be the chosen card.

As before, there is a slender chance that he will call out 1 as his number. Turn the top card face up. Say, "I thought so," and go on to another trick.

33   FUTURE THOUGHT

Numerology is the study of numbers and how they interact with our lives. It is generally agreed that numbers seem to have an influence not completely understood. It is because of this that evidence of the influence of numbers is indirect and frequently circumstantial.

As an illustration, someone jots down the first three digits in the phone number of a close friend. These three digits represent just the exchange of the given number. To impress the digits on his mind the spectator is asked to write these same three digits alongside the first set. Thus if he wrote 212 originally, the final number would be 212,212. The mentalist does not see the number.

Two dice are placed on the table. They are made from cardboard and they bear the numbers in Figure 25. The mentalist explains that a die is somewhat like a crystal ball in that it aids concentration. Each die is turned over a few times. Finally the mentalist pushes one die forward with the number 77 uppermost on the die.

"Obviously you didn't choose the number 77," the mentalist says. "But I caught a telepathic glimpse of the number you chose and I think that 77 will divide evenly into your six-digit number."

The spectator divides 77 into the number 212,212 and finds that it does indeed divide evenly into his number. Further, in most cases

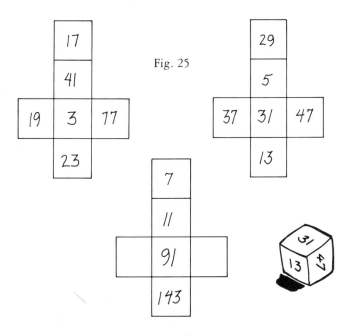

Fig. 25

Fig. 26

no other number on the die will exactly divide into the chosen six-digit number.

For the second test the mentalist says that he thinks he is on the same ESP wavelength as the spectator and would like to try a more difficult test, this time a prediction. He hides the other die under an inverted cup. The spectator now jots down another telephone exchange. As before, it is written twice. If the exchange this time is 791, the new six-digit number would be 791,791. The cup is lifted. The uppermost number on the die is 13. It is found that this number divides evenly into 791,791, and further that no other number on the die will evenly divide into the spectator's number.

As a final test the mentalist hands the spectator the die of Figure 26. This die has numbers on four sides but no numbers on two sides. "The blank faces are to help you concentrate," the mentalist says. "Gaze at one of the blank faces, then follow your instinct and turn the die so that one of the numbers is uppermost."

The spectator concentrates, then turns the die so the number 143, for example, is uppermost.

He is asked to jot down the telephone exchange of another friend and repeat the number to form a new six-digit number. The new exchange might be 934. When it is repeated the number becomes 934,934. He divides 143 into this number and finds that it divides exactly.

METHOD: This is L. Vosburgh Lyons' fine handling of an idea of Royal Heath's. After the dice are made up as described above, there is nothing to do except follow the routine exactly as written. The first time turn up the number 77 on one die. The second time turn up the number 13 on the other die. In the final phase of the routine the spectator chooses any number on the third die. This number will evenly divide into his chosen number.

Of course there is a chance that some other number on either die will divide evenly into the number chosen by the spectator but this only adds to the mystery. The point to remember is that 77 and 13 will *always* divide evenly into the numbers chosen by the spectator.

The subtle throw-off is that after each effect you allow the spectator to verify that most or all other numbers on either die will not work. Then you remove the third die from the pocket and have the spectator choose a number. All the numbers on this die will work, but he assumes that only the number he chose will divide evenly into his chosen number. It is this which makes the trick an unfathomable mystery.

34 THE DECODER CARD TRICK

This outstanding trick was shown to me by Howard Wurst, who learned it when he started in magic. Like the opening trick in this chapter, effect and method are intertwined and, also like that trick, it is one that is partly a card trick and partly a number trick. Children in particular are fascinated by the way the numbers are decoded to spell out the name of a chosen card.

It is necessary to memorize the number 543,235,621. If you have trouble keeping the number in mind, jot it down on a piece of paper and keep it handy. Refer to it just before you do the trick to refresh your memory. The number can be written in pencil on the outside of the cardbox and referred to when you remove the deck from the case.

Prepare by placing the Ace of Hearts on top of the deck. You are going to force this card on the spectator by a means known as the Cut Force or X Force. Place the deck on the table. Have him cut off about half the deck and place it alongside the bottom half, Figure 27. You then pick up the bottom half and place it on top of the other half, but crosswise, as shown in Figure 28.

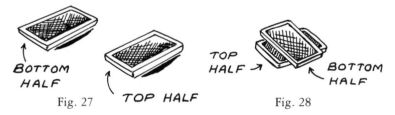

Fig. 27   *TOP HALF*   Fig. 28

Tell the spectator that you've developed a way of decoding random numbers so that they form words. Have him jot down a nine-digit number on a piece of paper. Under it you jot down the 9 complement—a number that, when added to the digit above it, will total 9. If the spectator writes 437,218,946, you would write 562,781,053 under it. This number is the 9 complement of his number. The two numbers would look like this when written on the paper:

$$4\ 3\ 7\ 2\ 1\ 8\ 9\ 4\ 6$$
$$5\ 6\ 2\ 7\ 8\ 1\ 0\ 5\ 3$$

Have him jot down another nine-digit number under your number. After he does this, you jot down a number that is the 9

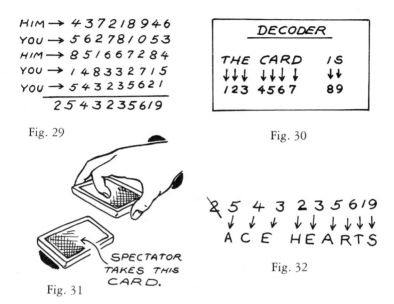

HIM → 437218946
YOU → 562781053
HIM → 851667284
YOU → 148332715
YOU → 543235621
——————————
2543235619

Fig. 29

DECODER

THE CARD   IS
↓↓↓ ↓↓↓↓ ↓   ↓↓
123 4567   89

Fig. 30

SPECTATOR TAKES THIS CARD.

Fig. 31

2 5 4 3 2 3 5 6 19
↓ ↓ ↓   ↓↓ ↓↓↓
A C E HEARTS

Fig. 32

complement of his second number. After you have written your second number you then jot down the memorized number, 543,235,621. Have the spectator add up all five numbers. The result is shown in Figure 29.

On another piece of paper write "The Card Is" and under it the digits from 1 to 9 in order, each digit under a letter, as shown in Figure 30. Copy his total on this piece of paper. Cross out the first digit on the left, remarking that your decoder only covers nine digits so you want to use just the last nine digits of his total.

Say, "Before we begin decoding, let's see what card you cut to." Lift up the upper packet in Figure 31 and have the spectator remove the top card of the packet still on the table. Although the card appears to be a random selection, it will be the Ace of Hearts. Pretend you don't know what it is.

Referring now to the spectator's total, after crossing out the first digit (a 2), the next digit is 5. According to the decoder chart of Figure 30, 5 is the same as the letter A. Jot this down under the 5 in the total. The next digit is 4. According to the decoder, 4 is equivalent to the letter C. Jot down this letter under the 4 in the total. Continue in this way, converting each digit to a letter. The amazing result is shown in Figure 32. The letters spell out the card chosen by the spectator.

MAGIC SQUARES

Magic squares are among the best-known mathematical diversions and though they are of venerable age, they still exert a hypnotic fascination. A magic square is an array of numbers where each column, row, and diagonal add to the same number. An example of a magic square is shown in Figure 33A.

L. Vosburgh Lyons once related the story of a performance given by Royal Heath for a large audience. Standing at a blackboard onstage, Heath started to construct a giant magic square. When he had almost finished filling out the square, Heath realized he had made a mistake. He erased the blackboard, then started again from the beginning. Although one might expect the audience would have become restless in the interim, they were quiet and attentive throughout. When Heath completed the magic square he received a rousing ovation.

The material in this chapter can be performed for audiences of 1 to 1000. There are no intricate formulas or complex calculations. The idea is to present the reader with ideas that are novel and entertaining. Try these demonstrations just once and you will be convinced of their value.

The chapter opens with "The Marrakech Game," an offbeat game which would appear to have no relation to magic squares. The surprising connection between Marrakech and magic squares is explained, and following this the reader will find a strange variation on the magic square which uses a fragment of a square. Methods of constructing magic squares can be found in Royal V. Heath's *Mathemagic* (Dover 20110-4). In this chapter we will deal with a streamlined version not explained in the Heath book. It is called "The Automatic Magic Square" (No. 42) and it is the routine that closes this chapter.

In this chapter the emphasis is on tricks that can be learned quickly, are easy to do, and have a strong impact on the audience.

35 THE MARRAKECH GAME

Call a friend and tell him that you've learned a new betting game. Have him bring a pencil and paper to the phone and jot down the numbers from 1 to 9. Tell him that you will take turns calling out numbers. He is to circle his numbers and put a square around the numbers you call out. The first person to get three numbers that add to 15 is the winner.

You go first and you call out 8. He might call out 6. You then call out 2. He calls out 5, you call out 4. Then he calls out 7 and you call out 3. At this point you announce that you have won because you called out 8, 3, and 4, which add to 15.

The spectator will want to play the game again. You are happy to oblige because although the game seems complex and beyond analysis, you cannot lose.

METHOD: Although the game *seems* tangled and obscure, there is a simple system which reduces it to childish simplicity. There will be games which you may play to a draw, but once the system is understood, there will never be a game you lose. You might respond that this is all well and good, but there seems to be no reason why the game is included in a chapter on magic squares.

Actually the game *is* a magic square, but perhaps even more unexpected, "The Marrakech Game" is really nothing more than tic-tac-toe.

To see how these seemingly disparate elements link together, first note the magic square shown in Figure 33A. The numbers in each horizontal row, each vertical column, and each major diagonal add to 15. For example, the digits in the diagonal that goes from the upper left corner to the lower right corner are 8, 5, 2. These digits add to 15. The digits on the other diagonal are 6, 5, 4. These digits also add to 15.

Now consider the game of tic-tac-toe. Suppose you played the game in an array that was numbered like a magic square. The start of one such game is shown in Figure 33B.

To say that the winner has to get three X's in a row or three O's in a row is the same thing as saying that he must get three numbers that will add to 15. In the example of Figure 33B the player with the X's will win if he plays an X in the box marked 4. He will get three X's in a row, but he will also get three numbers (8, 3, 4) which add to 15.

Now let's return to "The Marrakech Game." When you call a

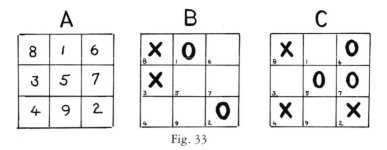

Fig. 33

friend, you have an array like that of Figure 33A in front of you. The first number you call is 8. Place an X in the 8-box of your array. He calls 6. Place an O in the 6-box. Your next number is 2 and his next number is 5. Your next number is 4 and his is 7. The result to this point is shown in Figure 33C. Which number should you call out next in order to have the digits add to 15? Looking at Figure 33C, it is easy to see that if you place an X in the 3-box, you will win. Not only have you gotten three X's in a row, you have gotten three numbers (8, 3, 4) which add to 15.

If you play "The Marrakech Game" using the magic square to record each player's turn, you cannot lose. If each player made his best move the game would end in a draw, but the spectator you call on the phone has no idea how simple the game is. He is trying to juggle partial totals while you calmly and rationally call out numbers that inevitably produce a strong, aggressive game. Even if he goes first you cannot do worse than draw. Most often you will win.

36   THE BROKEN SQUARE

This curious version of the magic square was demonstrated by Stephen Clark. The story is told that a fragment of an ancient papyrus was found. The incomplete square is shown in Figure 34.

The magician jots something down on a slip of paper. Then the spectator is asked to choose any row, column, or diagonal and to add the digits. When the spectator adds the digits, he might arrive at a total of 30. The slip of paper is opened and it correctly predicts that 30 would be chosen.

METHOD: The trick works the same way as any magic square. Any row, column, or diagonal adds to 30. But to those familiar with more traditional forms of the magic square, the fragmented square of Figure 34 seems incomplete and therefore unworkable.

	7	12	11
13	8	1	8
3	4	17	6
14	11		5

Fig. 34

Another way to present the trick is to draw up the square as shown, then tear out two squares, jot down the numbers 3 and 0 as indicated in Figure 34, and remark as you do so that you once received a magic square from a mystic but the square was in pieces. Place the two loose squares aside, writing-side down, as if they weren't important.

Have the spectator choose any row, column, or diagonal and add the numbers. When he arrives at a total of 30, remark that the total looks familiar. Turn over the two loose pieces and form the number 30, saying that this is the first time the total of a row was equal to part of a row.

37 SWINDLE SQUARE

Sometimes the most baffling tricks turn out to be the easiest to perform. In this trick you have no control over the numbers to be entered in a magic square, yet all columns and rows still add to the same total. There are many ways of doing a trick of this kind, but none is easier than the version described here.

Hand out slips of paper to each of six spectators. On each slip you've written a number. Explain that you'd like the first spectator to multiply his number by 1, the second to multiply his number by 2, the third spectator to multiply his number by 3, and so on up to 6.

While this is being done, draw a 6 × 6 array of empty squares. Then tell any spectator to stand and read off the product he got when he multiplied his number by one of the digits from 1 to 6. Remember that *any* spectator can stand. You enter whatever number he reads off in the uppermost row of boxes in the array as shown in Figure 35.

Another spectator rises and calls out his number. Whatever it is, enter it in the second row of the array. Continue with each of the other spectators. They can stand in any order they choose. Each person's number is different. You do not fake the writing; each number is entered into the array just as it is given to you.

Remarkably enough, when the array is completed as shown in Figure 35, each row and each column adds to 27.

METHOD: Although the trick seems incomprehensible to those not in the know, the secret is simple. The number you place on each slip of paper is 142,857. When this number is multiplied by any digit from 1 to 6 the result is the same six digits but in

7	1	4	2	8	5
1	4	2	8	5	7
5	7	1	4	2	8
8	5	7	1	4	2
4	2	8	5	7	1
2	8	5	7	1	4

Fig. 35

scrambled order. The only possible outcomes are those shown in Figure 35. It makes no difference how they are entered in the array. All rows and columns will add to 27. The diagonals don't sum to 27 but you avoid mentioning this point. Should someone begin to total the numbers on the diagonals, put the paper away.

38 ANTI-MAGIC SQUARE

There are many ways to number a square so that it has the "magic" property that all rows, columns, and diagonals add to the same number.

But suppose you want to find a square that gave as many *different* totals as possible. With a 3 × 3 magic square, how many different totals can be produced?

5	1	3
4	2	6
8	7	9

Fig. 36

The answer is shown in Figure 36, where it will be found that eight different totals are produced. This magic square is a good one to have in reserve for the fellow who knows the traditional form of the magic square. When asked to construct an anti-magic square, he may think it easy but the solution is not at all obvious.

39 CURIOUS MAGIC

After the spectator has struggled to complete the anti-magic square, you are ready to spring on him a rather different problem. He is to construct a 4 × 4 magic square such that it adds to 264 in all directions. This in itself is not difficult but there is an added stipulation. Not only must all numbers be different, but the magic square has to add to 264 in all directions *even when the square is turned upside down.*

The solution described by Heath is shown in Figure 37. When filling out the array, make sure the numbers can be read upside down. In other words the number 89 becomes 68 when the square is turned upside down, the number 91 becomes 16, and so on.

96	11	89	68
88	69	91	16
61	86	18	99
19	98	66	81

Fig. 37

40 ADD AND SUBTRACT

You don't have to be a genius to solve this problem, but a fair amount of ingenuity would be required to solve it quickly. It appears childishly easy right up to the point where you try to solve it.

The problem is this. The numbers 1 through 8 are to be paired with the numbers 9 through 16 so that each number in the first group is paired with a number in the second group. When the numbers have been paired, first add the numbers in each pair and then subtract one from the other. Thus each pair produces two other numbers. The stipulation is that the numbers must be paired in such a way that the 16 results obtained by the add-subtract process must all be different.

For example, you might choose to pair 2 with 9. With the add-subtract operation you would generate the sum $9 + 2 = 11$, and the difference $9 - 2 = 7$. But if you then decide to pair 6 with 13, you run into a problem. $13 + 6 = 19$, but $13 - 6 = 7$, and 7 is a result we have already obtained when we subtracted 2 from 9.

So the question is, how are the numbers paired such that all of the add-subtract operations produce different results? And what is the relation between these numbers and a magic square?

Each question should prove difficult to answer, even for the expert at magic squares. There are several solutions. L. Vosburgh Lyons suggested the following solution:

9/3 10/6 11/8 12/1 13/4 14/7 15/5 16/2

Felix Greenfield suggested the following solution:

9/2 10/5 11/7 12/4 13/1 14/8 15/6 16/3

The relationship to magic squares is that the numbers in either solution (or any other of the several solutions) can be arranged to form a semi-magic square, such that any horizontal row or vertical column, either major diagonal or any square block of four cells will add to within 1 of 50. The semi-magic square of Figure 38 uses the sums and differences of the Lyons solution. The square of Figure 39 uses the sums and differences of the Greenfield solution.

19	3	17	11
16	12	18	4
9	14	6	21
7	20	10	13

Fig. 38

19	4	16	11
15	12	18	5
8	14	6	22
7	21	9	13

Fig. 39

41 THE PARADOX

An ingenious paradox is worked into this magic-square routine. The paradox might succinctly be described as a five-piece puzzle that is constructed with only four pieces.

The routine is as follows. Using five pieces of paper the magician forms the magic square shown in Figure 40. Each row, column, and diagonal sums to 45. The five pieces are turned over. Now the spectator forms the magic square indicated on this side of the paper, Figure 41. Strangely enough he requires only four pieces to form this magic square. He sees that the sum is 15 in each direction. Turning over the piece he did not need, he finds it bears the number 15, exactly matching the sum of the square, Figure 41.

METHOD: Only a few minutes are required to construct the Paradox Square. Use a sheet of paper measuring 8½″ on a side. Mark off a point 3″ from each corner and connect these points with light pencil lines, as indicated by the dotted lines in Figure 42. You will also need a square piece of paper measuring 2½″ on a side as in Figure 42.

Draw the numbers shown in Figure 43, using black pencil, crayon, or marking pen. Note that the numbers are drawn on an angle, parallel to the dotted lines. Draw the number 15 on the small square of paper shown in Figure 43. Again note that the number is drawn on an angle.

Now turn the paper over side for side (not end for end). Draw the numbers shown in Figure 44 with red pencil or marking pen. The numbers must be drawn exactly as shown.

Turn the paper over and cut it out along the dotted lines. With the apparatus in hand you are able to perform the trick. Place the pieces on the table with the black numbers up. There are the four pieces you just cut out plus the square piece with the number 15 on it.

Form the magic square as shown in Figure 45. Note that it takes all five pieces to form the square. Let the spectator verify that the sum is 45 in each direction. Then gather all five pieces and turn them over so the red side is uppermost. Let the spectator form the magic square. He will be baffled that he formed the square with only four pieces, Figure 46. The magic square goes together only one way, so he must end up with the correct square.

Let him verify that the total is 15 in all directions. Then have him turn up the square piece he didn't use, Figure 46, to see that this piece reveals the sum of the square he just constructed.

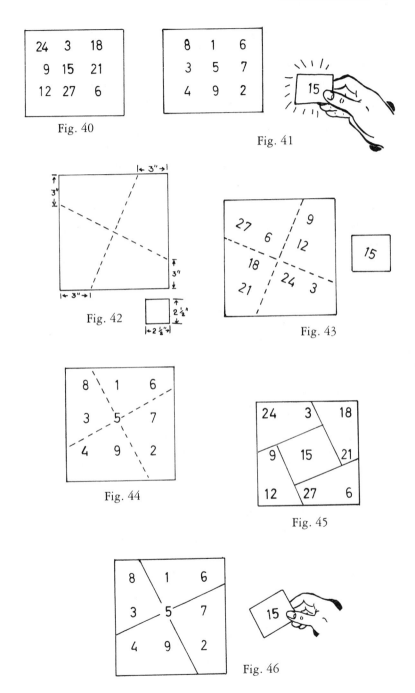

Fig. 40

Fig. 41

Fig. 42

Fig. 43

Fig. 44

Fig. 45

Fig. 46

42   THE AUTOMATIC MAGIC SQUARE

This chapter closes with an astounding demonstration of mental powers. You can present it as a feat of lightning calculation or as a mental effect in which you read the spectator's mind. Either way it is one of the most incredible demonstrations with magic squares.

The effect is this. The spectator calls out any number between 22 and 100. Assume he calls out 53. Immediately you fill out a 4 × 4 magic square such that every row, every column, and each diagonal adds to 53, as shown in Figure 47.

Further, each group of four corner squares (Figure 48), the outermost corner squares (Figure 49), and even the center four squares (Figure 50) add to 53. Remember that you fill in the magic square *instantly*. There is no hesitation. The instant the spectator calls out the number, you begin writing, and you fill in the squares as fast as you can write.

Fig. 47

Fig. 48

Fig. 49

Fig. 50

METHOD: The demonstration would appear to require enormous powers of mental calculation but it is in fact much easier than it looks. The secret is that you begin with a square that is almost complete. The basic array is shown in Figure 51. Note that all the cells except A, B, C, and D are filled in. These four squares are the only ones that concern you.

Ⓑ	1	12	7
11	8	Ⓐ	2
5	10	3	Ⓓ
4	Ⓒ	6	9

Fig. 51

When the spectator calls out a number, say 53, mentally subtract 21 from it. Whatever the result, put that number in square A. In our example you would put 32 in square A. Add 1 to the number to get 33 and put it in square B, add one to that, putting 34 in square C, and, likewise, 35 in square D. In other words, you place 32, 33, 34, 35 in squares A, B, C, D respectively. Since this is all you need do, you can see that it can be accomplished as quickly as you can write.

If you perform "The Automatic Magic Square" frequently, you will be able to recall the array from memory. If you do the stunt only once in a while, jot down the array of Figure 51 on a slip of paper and carry it in your wallet. Consult the paper to refresh your memory just before doing the stunt. Of course A, B, C, and D are left blank when you draw the magic square and fill in the individual numbers. It is a simple matter to remember the positions of A, B, C, and D.

Should you use a blackboard, the numbers can be written in pencil since pencil writing is not visible to the audience. When it comes to filling in the square, write in chalk directly over the pencil writing. If paper is used, Don Nielsen suggests that the numbers be typed but with the typewriter ribbon removed. A white-on-white impression will be on the paper, visible to you but not to anyone else.

2	1	12	7
11	8	1	2
5	10	3	4
4	3	6	9

Fig. 52

Before getting to the presentation there is one point that should be mentioned. The spectator is given a choice of numbers from 22 to 100. If he calls out a number between 22 and 34, you will find that some of the numbers will be duplicated. For example, if he calls out 22, the complete magic square will look like Figure 52. Note that 1's, 2's, 3's, and 4's are duplicated in the array. It is a small point and is not likely to be noticed in view of the enormity of the accomplishment, but if you are bothered by the duplication of some digits, limit the spectator's choice to any number between 34 and 100.

In presenting "The Automatic Magic Square" you may prefer an approach that emphasizes ESP ability. The spectator is asked to think of a number between 22 and 100. While he concentrates on the number, take out the paper and pencil. Close your eyes and act as if you are focusing on supernatural wavelengths. Draw the 4 × 4 array, then place the digit 1 in its proper place. Pause again, then put the 2 in its proper box. Fill in three or four more digits. The spectator knows you're writing something but he doesn't know what. You can continue filling in the numbers up to 12.

Say, "I think I've got it. What number were you thinking of?" As soon as he tells you, subtract 21 from the number and fill in squares A, B, C, D, acting as if you are making a few trivial notes. Nod your head and place the paper writing-side down on the table.

Say, "Are you familiar with magic squares? That's where the numbers add up to the same total. Like this." Turn over the paper and have the spectator add up the row, columns, and diagonals. Each time, the total is the same as the number he was thinking of. Obviously you must have known the number because you filled in the magic square *before* he told you the number.

Performed this way the trick is more of a mind-reading trick than a stunt of the lightning calculator. If you want to emphasize the mental gymnastics, start with a blank array. As soon as he calls out his number, subtract 21. This gives you the number that goes in A, but don't put the number there. Instead add 1 to it and put this number in B. Then fill in the rest of the top row from left to right. Without hesitation fill in the second row from left to right, then each of the other two rows.

Try it each way and choose the approach that works best for you. Regardless of your choice, you will find it an exceptional routine.

THE CLASSICS

Not all stunts with numbers require serious presentation. Some of the best puzzles and bets with numbers have a whimsical quality. Appearing clear and logical on the surface, they seem to defy analysis.

The puzzles in this chapter represent some of the classics along with new material. Interspersed with the more serious mysteries described elsewhere in this book, they cannot help but enhance your reputation as a number expert.

43 THE ROOT OF THE PROBLEM

During a classroom discussion on square roots Johnny wrote the equation of Figure 53 on the blackboard. It was obviously wrong and the teacher was about to cross it out, but on second thought it seemed *almost* right. Is the equation correct? Almost correct? Completely wrong?

(All answers are at the end of this chapter, pages 81–86.)

$$2\sqrt{\frac{2}{3}} = \sqrt{2\frac{2}{3}}$$

Fig. 53

44 HAUNTED HOTEL

Three guests checking into the hotel were told that rooms were $15 apiece. They each gave the bellhop $15, a total of $45. When the desk clerk heard about this he reminded the bellhop that rooms were only $10 apiece. He kept $30 for the room and told the bellhop to return the balance. On the way upstairs the bellhop reasoned that since the guests didn't know exactly how much the rooms cost, they would be happy with any rebate. The bellhop gave each of the three guests $3, a total of $9, keeping $6 for himself.

Here is where the problem arises. Each guest paid $12 for his room, a total of $36. The bellhop kept $6. We have thus far accounted for $42. What happened to the other $3?

45   ARITHMETRICK

Certain number problems have a slippery logic about them. The problem is simple enough, but the solution seems to lie just out of reach. An excellent example of this type of problem is the following. A fisherman announces that he caught a fish which had a length of 30 inches plus half of its own length. How long is the fish?

The answer seems simple enough. Half of 30 is 15, so the fish was 30 + 15 or 45 inches long. But this approach assumes that the fish's length is 30 inches and that we take half of its length to find the actual length. Clearly the length can't be both 30 inches and 45 inches.

Returning to the original problem, we now seem to be at an impasse since the wording yields no new information. Indeed, it appears that a contradiction is involved since a fish can't be equal to something plus half its *own* length. At this point we might conclude that it's all just another fish story and that the fisherman didn't catch such a fish in the first place, but he did, and the original question remains: how long was the fish?

46   CONTRARY CLOCK

About to leave his office for home, John notices that the clock in the office shows the wrong time. Since he doesn't have a watch of his own, John walks home, has a snack, notes the correct time, then walks back to the office and sets the office clock at the correct time. Assuming he walks at a constant pace and takes the same path home and back to the office, how did he know how to set the office clock to the right time?

If one knew how fast he walked and how far it was between the office and home, it would be easy to calculate the time for the round trip. The odd thing is that this information is not required to solve the problem.

47 MOUNTAIN TIME

If someone happens to solve the above problem, this follow-up is good because it seems to use the same logic. In fact it is quite different.

Arriving at their cabin at the base of the hill, newlyweds Jack and Jill discover that the cabin clock shows the wrong time. There is a clock at the top of the hill near the well and they know it is always correct. Jack walks to the top of the hill to fetch a pail of water and sees that the well clock says 12 noon exactly. He walks back down the hill to his cabin. Can he now set the cabin clock to the correct time?

The answer may seem to follow from the reasoning used in "Contrary Clock" but note here you can't assume he walked up and down the hill at a constant rate. Obviously it must take him longer to climb up the hill than to climb down. Since we don't know how long it took him to climb the hill, we must conclude that he can't set the cabin clock at the right time.

The curious point about this problem is that if he decides to reclimb the hill, he will be able to predict the exact setting of the well clock when he arrives at the top of the hill.

He does it by this method. Suppose the cabin clock said 3:00 when he left to climb the hill. When he gets to the top of the hill he notes that the well clock reads 12 noon. When he returns to the cabin at the foot of the hill, the cabin clock now reads 4:00. Thus he knows that the round trip (forgetting the matter of fetching a pail of water) takes an hour.

If he now climbs the hill again, he knows he will get to the well clock when that clock shows 1:00. A further curious point is that he can also predict the setting on the cabin clock when he returns. If he left with the cabin clock showing 4:00, it will show 5:00 when he returns.

The odd problem facing Jack is that he can predict with perfect accuracy his time of arrival at the top or bottom of the hill, yet he is unable to set the cabin clock at the right time. There is a simple way around the problem, though. Can the reader find it?

48 SIGNIFICANT DIGITS

It is possible to give an instruction to someone such that he will write the number 541,632 without you or anyone else mentioning any of the digits. The spectator is not a confederate, yet he will infallibly write the six-digit number after hearing a simple instruction from you. Remember that your instruction cannot contain any numbers. What is the instruction?

49 UNWRITTEN RULE

A certain rule book, which lists the rule for conjuring up demons, contains thousands of rules, but some are difficult to find. By way of example, ask the spectator to write the number for rule nine thousand nine hundred and nine. He will of course write 9909. Then ask him to write the number for rule twelve thousand twelve hundred and twelve. Surprisingly, he will probably not be able to write the number correctly.

50 UNEQUALS

The equation shown here may appear to be wrong. What punctuation marks should you add to make it correct?

$$560 = 600$$

51 WHAT'S NEXT?

Given the series of letters O, T, T, F, F, S, S, what is the next letter in the series? If you know the answer, then you can immediately state the letter that is ninety-fifth in the series.

Given the series of letters T, D, D, H, H, H, what letter is next in the series? What letter is ninety-fifth in the series?

This problem is a bit trickier. Given the series of letters T, H, T, T, T, H, T, what letter is next? You might think at first that the H's and T's stand for Heads and Tails, and you may be right, but if so, can you state positively what the next letter is going to be?

52   CRYPTARITHMS

A cryptarithm is a problem in which letters replace numbers. You are then to figure out what the numbers are supposed to be. If the same letter shows up more than once it is assumed to have the same numerical value. A sample is this:

$$
\begin{array}{r}
A\ B \\
+\,B\ A \\
\hline
A\ A
\end{array}
$$

It can be seen at a glance that B must be zero. There is not enough information to determine a value for A, but if you were further told that:

$$
\begin{array}{r}
A\ A \\
+\quad A \\
\hline
C\ B
\end{array}
$$

the solution falls into place. We know already that B = 0. If AA + A adds to a number that ends in zero, A must be 5, because 5 + 5 = a number that ends in zero and it is the only digit that has this property. All that remains is to specify that C = 6.

If you feel you understand the subject you may want to tackle a classic problem suggested by Henry Dudeney:

$$
\begin{array}{r}
A\ B \\
\times\quad C \\
\hline
D\ E \\
+\,F\ G \\
\hline
H\ I
\end{array}
$$

The problem may appear complex until you realize that what Dudeney has done is to arrange the digits from 1 to 9 in an array where each digit is used once. The solution should then make itself readily apparent.

53   EVE DID TALK

This is one of the best cryptarithms, a combination of logical precision and amusing commentary:

$$
\frac{EVE}{DID} = .TALKTALKTALKTALKTALKTALK \ldots
$$

54   ORDERLY TRANSITION

Given the digits 1, 2, 3, 4, 5, 6, 7, 8, 9 in order, insert plus and minus signs such that the numbers will total 100. One solution, using just three signs, is this:

$$123 - 45 - 67 + 89 = 100$$

You may now wish to consider the opposite problem. Given the digits in reverse order, i.e., 9, 8, 7, 6, 5, 4, 3, 2, 1, insert plus and minus signs so that the numbers will total 100. Can you solve the problem with just four signs?

55   NAMING NUMBERS

Fill in the boxes of Figure 54 with digits such that the digit in the first box is the same as the number of 0's in the number, the digit in the second box is the same as the number of 1's in the entire number, the digit in the third box indicates the number of 2's in the number, and so on.

Fig. 54

56   ZEBRA ALGEBRA

A zebra clocked at 40 miles per hour took 80 minutes to run through the jungle from east to west. But when he ran the same level path through the jungle from west to east at the same speed, it took him an hour and 20 minutes. How could this be?

57   FIDGET DIGITS

This seemingly simple problem was demonstrated by L. Vosburgh Lyons. It contains a little-known kicker that appears in print here for the first time. The problem is to take the digits 1 through 8 and place them in the array of Figure 55, one digit in each square, such that no two digits that are next to each other in counting order are next to each other in the array. Thus 4 and 5 can't be adjacent to one another either horizontally, vertically, or diagonally; 5 and 6 can't be adjacent to one another; 6 and 7 can't be adjacent; and so on.

The problem may be more difficult than it first appears. If the spectator tries and gives up, show him the solution given at the end of this chapter.

Then comes the kicker. Place the solution in your pocket, take out another piece of paper and ask the spectator to draw the solution from memory. As you speak, draw the array shown in Figure 56. It looks the same as the original array but it is just different enough to confuse the spectator.

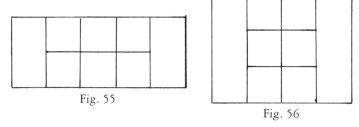

Fig. 55

Fig. 56

58 STRANGE SYMBOLS

While wandering around in Looking-Glass Land, Alice came upon the strange symbols shown in Figure 57. At first the symbols made no sense, but after studying them a moment Alice was suddenly able to understand what they meant. Can the reader decipher the meaning?

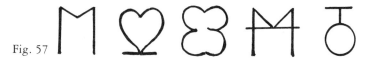

Fig. 57

59 CATCH QUESTIONS

If there are 52 cards in the deck and the gambler took all but two for himself, how many cards would be left for you? Oddly enough, the answer is not 50.

If the gambler divided the 52-card deck by one-half and added three Jokers, how many cards would be in the deck?

60 SCRAMBLED STATES

Certain of the 50 states contain numbers. For example, which of the 50 states begins with 10? Tennessee is a state that begins with TEN but that is not the answer to the question. Another question involving states and numbers is, which state ends with 10?

61 PUZZLE PARTS

The digits from 1 through 9 can be divided into two parts as follows:

Part One = 1, 4, 7   Part Two = 2, 3, 5, 6, 8, 9

If you are told that 10 fits into Part Two and 11 into Part One, into which part would 12 fit? How about 88?

62 SIX QUIK

Circle six different digits in the group shown in Figure 58 such that when added up they total 22. You should be able to do it with just a little thought. Then try your hand at circling six digits so they total 21 or 24. You may get one of these totals easily but the other will give you trouble.

$$
\begin{array}{ccc}
1 & 1 & 1 \\
3 & 3 & 3 \\
5 & 5 & 5 \\
9 & 9 & 9 \\
\end{array}
$$

Fig. 58

63 WHO PAYS?

There's a town south of the United States—Mexican border where a U.S. dollar is worth 90¢ in Mexican currency. In the town just north of the border a Mexican dollar is worth 90¢ in U.S. currency.

A cowboy went into the Mexican town and bought a 10¢ soda. He paid for it with a Mexican dollar, and he was given a U.S. dollar in change since the American dollar is worth only 90¢ there.

Crossing the border into the U.S., he bought a 10¢ soda in the American town, paid for it with the U.S. dollar he just got in Mexico, and received a Mexican dollar in change since the Mexican dollar is worth only 90¢ north of the border.

He goes back to Mexico and buys another soda, then back to the U.S. and buys another, and does this day and night, always ending up with a dollar, just as he started with. Who pays for the soda?

64   THE MONKEY AND THE COCONUTS

The story is told of five sailors and a monkey being shipwrecked on an island. The men spent the day gathering coconuts into a heap. Then they went to sleep. In the middle of the night one of the men woke up and thought it might be wise to take his fair portion of the coconuts. He divided the coconuts into five equal piles. One coconut was left over so he tossed it to the monkey. He took his portion and hid it. Then he went back to sleep.

A second sailor woke up, divided the remaining coconuts in five equal piles, and noticed there was one left over, so he threw it to the monkey. Then he took his pile, hid it and went back to sleep.

The third, fourth, and fifth sailor each woke up in turn and did exactly the same thing. At dawn all five sailors woke up and divided the remaining coconuts evenly. There were no coconuts left over for the monkey. The question is, how many coconuts did they originally gather together?

ANSWERS TO "THE CLASSICS"

All of the problems in this chapter are answered here. Although in some cases the analysis is complex and outside the scope of this book, the basic approach is sketched in so the reader knows the direction to take. The only exception is the final problem, "The Monkey and the Coconuts," which is far more tangled than it may at first appear.

The Root of the Problem
The problem shown in Figure 53 is correct as written.

Haunted Hotel

The trick here is that at the end of the problem you are asked to relate the amount spent by the men to the amount gained by the bellhop, a misleading question. To see why, assume you gave me $10. That means that you are out $10 and I have made $10, a total of $20. But if we started with only $10, where did the other $10 come from? You can see that there is no other $10.

The hotel problem conceals this with slick wording. The desk clerk has $30, the guests have $9, and the bellhop has $6, a total of $45, exactly what the problem started with.

Arithmetrick

If a fish's length is equal to a certain amount plus half its own length, then the certain amount must be the *other* half. Now the problem falls into place. The fish in question must be 60 inches long.

To say it another way, if a fish is X inches long plus half its own length, to find its length we need only take twice X and we have it. If for example a fish is said to be 90 inches long plus half its own length, we know now that the fish must be 180 inches long.

Contrary Clock

Before leaving the office John noted the time on the office clock. Then he went home and noted the time on the kitchen clock. He had a snack, noted the time on the kitchen clock, and went back to the office.

Since he knew the time on the kitchen clock when he got home, and also the time when he left to return to the office, he knew how much time he took to have a snack. Let's say snack time was a half-hour.

When he got back to the office he saw that 50 minutes had elapsed from the time he left. Deducting 30 minutes snack time from 50 minutes total elapsed time, we get 20 minutes walking time. That is, it took him 20 minutes to make the round trip.

Half of this is 10. Thus we know that it took him 10 minutes to walk each way. If he knew the time on the kitchen clock when he left home, he needs merely to set the office clock at this time plus 10 minutes. The office clock will now show the correct time.

Mountain Time

By prefacing this problem with the clock problem, and by referring only to Jack's excursions up and down the hill, the reader's attention has been directed away from the very simple solution because the reader is misdirected into thinking that Jack was alone.

Actually he was with his wife Jill. When he got to the top of he hill he hit the pail against the well to signal he just reached the well. Jill then noted the time on the cabin clock. Let's say it said 3:40.

Jack knew he left the cabin when the cabin clock said 3:00, so he now knows that it took him 40 minutes to reach the top of the hill. After he fills the pail with water he notes the time on the well clock and heads down the hill to the cabin. Knowing that the round trip took an hour, and that it took 40 minutes to get to the top of the hill, Jack knows it took 20 minutes to reach the bottom. All he has to do is set the cabin clock for the time at the well clock when he left, plus 20 minutes, and the cabin clock will now be at the correct time.

Significant Digits

This clever problem was suggested by Wallace Lee. The instruction to the spectator is to list the first half dozen digits in alphabetical order. To do it correctly he must write them in the order 541,632.

Unwritten Rule

The correct answer is 13,212.

Unequals

The answer is that 5:60 = 6:00 when telling time.

What's Next?

The first series is merely the first letter of the numbers One, Two, Three, Four, and so on. The ninety-fifth letter in the series is the first letter of Ninety-Five, or N.

The second series represents the last letter of the words First, Second, Third, Fourth, and so on. The next letter is H, as is the ninety-fifth letter in the series.

The third series looks as if it might stand for a progression of Heads and Tails, but the series is generated by taking the first letter of progressively larger multiples of Ten, i.e., Ten, Hundred, Thousand, Ten Thousand, Hundred Thousand. From this one can see that the next letter in the series must be the M in Million.

Cryptarithms

The solution to the Dudeney problem is

$$
\begin{array}{r}
1\,7 \\
\times \quad 4 \\
\hline
6\,8 \\
+\,2\,5 \\
\hline
9\,3
\end{array}
$$

Eve Did Talk

To find the fractional equivalent of a decimal number, put the basic cycle over an equal number of 9's. In this case we would get

$$.\text{TALK} = \frac{\text{TALK}}{9999}$$

Now the original puzzle reads as follows:

$$\frac{\text{EVE}}{\text{DID}} = \frac{\text{TALK}}{9999}$$

DID must be a factor of 9999. Trying different possibilities you will find that the solution must be

$$\frac{242}{303} = .79867986\ldots$$

Orderly Transition

The solution is $98 - 76 + 54 + 3 + 21 = 100$.

Naming Numbers

The solution is shown in Figure 59.

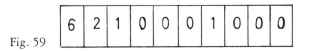

Fig. 59

Zebra Algebra

80 minutes is the same as 1 hour and 20 minutes.

Fidget Digits

The first solution is shown in Figure 60. The solution to the second part of the problem is shown in Figure 61.

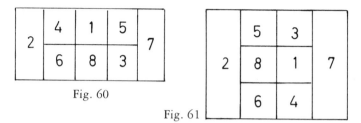

Fig. 60

Fig. 61

Strange Symbols

The problem contains a clue in the wording. Alice saw the symbols while in Looking-Glass Land. The symbols actually represent the numbers 1, 2, 3, 4, 5, plus their reflections, as shown in Figure 62.

Fig. 62 ⌐1 ⌐2 ⌐3 ⌐4 ⌐5

Catch Questions

The gambler left two cards for you in the first problem. In the second problem, note that dividing 52 by ½ gives you an answer of 104 (dividing 52 by 2 gives you an answer of 26). Adding 3 jokers, there would be 107 cards in the deck.

Scrambled States

The correct answer to the first question is IOWA. The answer to the second question is OHIO.

Puzzle Parts

The rule that decides what numbers go in which part is that numbers drawn with straight lines go into Part One. Numbers that contain curved lines go into Part Two. Thus 10 goes into Part Two, 11 into Part One, and 12 into Part Two. Since 88 consists of curved lines, it goes into Part Two. This is really a problem in pattern recognition, and is not obvious until it is pointed out.

Six Quik

This problem was suggested by Martin Gardner. You will have no trouble arriving at totals of 22 and 24, but you might think it impossible to find six digits that total 21, and not without reason. It is indeed impossible to construct an odd number from six odd digits.

This is a clue that there is a trick to it. The trick is that to form the total 21, you must first turn the group of numbers in Figure 58 upside down. Then the problem is easily solved. To form the other totals, you have to view the numbers right side up. The solutions are:

$$21 = 1 + 1 + 1 + 6 + 6 + 6$$
$$22 = 1 + 1 + 1 + 5 + 5 + 9$$
$$24 = 3 + 3 + 3 + 5 + 5 + 5$$

Who Pays?
As things stand, no one pays.

The Monkey and the Coconuts
The smallest number of coconuts that will solve the problem is 3121.

CALENDAR CONJURING

Calendars are everyday objects that lend themselves to a number of puzzling tricks and stunts. The mathematical properties of calendars have been extensively investigated, but in this chapter we will consider only the most streamlined means of producing commercial calendar magic. The chapter concludes with Martin Gardner's remarkable calendar prediction.

65 WHAT A WEEK!

Hand the spectator a calendar. Have him turn to any month. While your back is turned have him circle three consecutive days during any week. These can be the three days he likes best, the slowest days at work, the best days for a vacation, etc. He will circle three days in a horizontal row as shown in Figure 63.

S	M	T	W	T	F	S
		1	2	3	4	5
6	7	8	9	10	11	12
13	(14	15	16)	17	18	19
20	21	22	23	24	25	26
27	28	29	30			

Fig. 63

Tell him to add mentally the three numbers and tell you the total. Say the total is 45. Ask, "What significance do these days have for you?" He might say that they represent a three-day trip he took to the Coast. To this you respond, "Of course. You were gone the 14th, 15th, and 16th." And you're right.

METHOD: The trick depends on the fact that he is adding consecutive numbers. The principle works for any three consecutive numbers, but since the spectator is limited to calendar days he will choose numbers that are easily handled without the need for pencil and paper.

When the spectator announces the total of the three numbers, divide the total by 3. The result will be the second date of the three he circled. Once the center date is known, subtract 1 to get the first date, and add 1 to get the third date.

In presenting the trick note that you are not merely revealing three numbers. Rather, you are telling the spectator when he took the trip to the Coast.

66 BEST DAZE

Each person looks forward to a particular day of the week. For some it is Friday, when work or school are over. For some it is Sunday when they can rest. For a few it is Monday because they can't wait to get back to the office. While your back is turned the spectator calls out his favorite day. Assume he says Thursday. Tell him to pick a calendar page and circle three consecutive Thursdays. The result might be as shown in Figure 64.

He adds the circled numbers and calls out the total. Immediately you tell him which Thursdays he picked.

METHOD: When he calls out the total, divide it by 3. The result will be the middle Thursday of the three he circled. Subtract 7 from this number to get the first date, and add 7 to it to get the third date.

In our example he adds 10 + 17 + 24 to get 51. When he calls out this total, mentally divide 51 by 3 to get 17, the middle Thursday. The other Thursdays are obtained by adding 7 to 17 to get 24, and subtracting 7 from 17 to get 10.

S	M	T	W	T	F	S
		1	2	3	4	5
6	7	8	9	10	11	12
13	14	15	16	17	18	19
20	21	22	23	24	25	26
27	28	29	30			

Fig. 64

67   FOUR OF A MIND

If asked to repeat "Best Daze," say you'll make it harder. Have the spectator turn to another calendar page and circle four dates in a vertical column. The result might be as shown in Figure 65. He announces the total of the four numbers. Immediately you tell him the dates he circled.

METHOD: When he announces the total, subtract 14 from it and divide the result by 4. The number you get is the second date he circled. The first date he circled will be 7 less than the second date. The other dates will be 7 more and 14 more respectively.

In our example the spectator announces a total of 54. Subtract 14, getting 40. Divide 40 by 4, getting 10. Thus you know that the second Thursday is the 10th. The first Thursday he circled is the 7th, the third Thursday is the 17th, and the fourth Thursday is the 24th.

S	M	T	W	T	F	S
		1	2	3	4	5
6	7	8	9	10	11	12
13	14	15	16	17	18	19
20	21	22	23	24	25	26
27	28	29	30			

Fig. 65

68   DAY FIVE

In some calendar months a particular day of the week will appear five times. Have the spectator find such a page in the calendar. While you turn your back have him circle the first four days and put a square box around the fifth day as shown in Figure 66.

He calls out the total of the first four days. You immediately tell him the date of the fifth day.

METHOD: The calculation of "Four of a Mind" can be used, but a more streamlined method is this. Whatever the total, add 70 to it and divide by 4. The result will be the boxed number.

In the example of Figure 66 the spectator calls out a total of 50. Add 70 to get 120, divide by 4 to get 30, and announce that the boxed number was the 30th.

A sneaky approach to this problem is as follows. It seems at first to be vastly more difficult but in fact it is the easiest method thus

S	M	T	W	T	F	S
			1	2	3	4
5	6	7	8	9	10	11
12	13	14	15	16	17	18
19	20	21	22	23	24	25
26	27	28	29	30	31	

Fig. 66

S	M	T	W	T	F	S
			1	2	3	4	5
6	7	8	9	10	11	12
13	14	15	16	17	18	19
20	21	22	23	24	25	26
27	28	29	30	31		

Fig. 67

far considered. Have the spectator find a month that has five of one day. Tell him to circle the first two and last two days in this column. The result might look like Figure 67. He announces the total of these four numbers and you instantly tell him the date he didn't circle!

Whatever total he announces, simply divide it by 4 and that's it. In the example of Figure 67 he will announce a total of 64. Divide by 4 and you arrive at 16, the Wednesday he didn't circle.

69 THE SQUARE WEEK

The time will come when the work week will be reduced to just four days. Rather than work four consecutive days the worker will have his choice of four days in a square array. Ask the spectator to draw a square around four days. He might circle the four days shown in Figure 68.

He calls out the total of the four days and you tell him which dates he picked to work.

METHOD: Whatever total he gives you, divide it by 4 and then subtract 4 from the result. This number is the first date he circled.

S	M	T	W	T	F	S
		1	2	3	4	5
6	7	8	9	10	11	12
13	14	15	16	17	18	19
20	21	22	23	24	25	26
27	28	29	30	31		

Fig. 68

The second date must of course be 1 more than the first. The third date is 7 more than the first, and the last date is 1 more than that.

In our example he would call out a total of 40. Divide by 4, getting 10. Subtract 4 from 10, arriving at 6. The first date he circled is the 6th. The second date must be the 7th. The third date is 7 more than the first, so it has to be the 13th. The final date is one more than that or the 14th.

70   CROSS OUT

Have the spectator draw a cross around five numbers such as that shown in Figure 69. He adds the four outer numbers but not the center number. When he announces the total of the four numbers you instantly tell him the center date of the cross.

METHOD: Whatever total he announces, simply divide by 4 and that will be the center date of the cross.

For a variation, have the spectator draw an X around five numbers as shown in Figure 70. He totals the four outer numbers of the X but not the center number. When he announces the total, divide it by 4 and reveal the result as the center number of the X.

Fig. 69   Fig. 70

71   SECOND SATURDAY

Ask the spectator to open any calendar to any month and tell you the date of the second Saturday. He might say the 10th, whereupon you immediately tell him that the first of that month falls on a Thursday!

METHOD: This trick can be accomplished by involved calculations but there is a surprisingly simple shortcut which involves no arithmetic. The secret depends on two facts, one fairly well known and the other almost unknown. The more familiar fact is that the

15th of any month falls on the same day of the week as the first of the month. Thus if you know which day of the week the 15th falls on, you know which day of the week the month begins with.

To perform the trick without calculations we make use of the little-known fact that the second Saturday of most months falls between the 7th and the 15th. Knowing the date of the second Saturday it is easy to move ahead to the 15th, advancing a day at a time.

Suppose for example you were told that the second Saturday was the 14th. You would advance 1 to the 15th, mentally advancing from Saturday to Sunday. Knowing that the 15th is a Sunday, you know that the 1st must also be a Sunday.

Another example is this. Say you were told the second Saturday fell on the 10th. You know you must advance five days to reach the 15th. Advancing five days from Saturday, you arrive at Thursday. Since the 15th falls on a Thursday, the first day of the month must also fall on a Thursday.

72   DATE SENSE

Throughout history the methods of keeping time have varied, from the sundial to the hourglass to the calendar. This unusual trick, based on an idea of Walter Gibson's, ties in all three means of recording the passage of time.

Give the spectator a calendar page. Ask him to look at each week and circle a single day in that week which he thinks might have significance to him. When he's circled a number in each week, the calendar might look like Figure 71. Of course your back is turned and you do not see which numbers he circled.

Ask the spectator, "How many Sundays did you circle? How many Mondays? How many Tuesdays?" Continue with this process, asking how many of each day of the week he circled. In our example he will say he circled one Sunday, no Mondays, one Tuesday, one Wednesday, one Thursday and one Friday. He circled no Saturdays.

As he talks, you slide an hourglass around the circle of numbers shown in Figure 72, explaining that the circle represents the markings found on ancient sundials. When the sand runs out of the hourglass you will stop at whatever number the hourglass happens to be resting on.

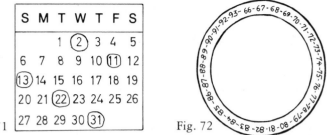

Fig. 71

Fig. 72

The number you stop on might be 79. The spectator totals his circled numbers and finds that the total is indeed 79.

METHOD: Try to use a calendar month that has five Wednesdays. The dial of Figure 72 is easily made by jotting down the numbers in a circle on a piece of cardboard. Inexpensive small hourglasses are available in department and hardware stores where they are sold as egg timers and decorative novelties.

First note the total of the five Wednesdays. In the example of Figure 71 the values of the five Wednesdays total 80. Place the hourglass at 80 on the dial when you begin the trick.

Have the spectator circle one day each week. You then ask him how many of each day was circled. The movement of the hourglass around the dial is governed by the following table:

Sunday	−3	Thursday	+1
Monday	−2	Friday	+2
Tuesday	−1	Saturday	+3
Wednesday	0		

A negative number means that you slide the hourglass in a counterclockwise direction. A positive number means that you slide the hourglass in a clockwise direction. For example, when you ask the spectator how many Sundays he circled, he will tell you that he circled one Sunday. In the table, Sunday = −3. Beginning with the hourglass on 80, move it three numbers in a counterclockwise direction. It will end up on the number 77.

He says no Mondays are circled, so you do not move the hourglass. He says one Tuesday, so you move the hourglass one number counterclockwise, to 76. Ignore Wednesdays, no matter how many he calls out.

He circled one Thursday, so you move the hourglass one space clockwise to 77. He circled one Friday so you advance the hourglass two spaces to 79.

Have him total the circled numbers. He will arrive at 79, exactly matching the position of the hourglass.

Once you become familiar with the trick, you can bluff by moving the hourglass a random number for Wednesdays or for days he didn't circle. Try to keep the hourglass in constant motion. Time the movement so that the sand runs out just as you slide into the correct number.

73 CRAZY CALENDAR

When the calendar was invented, it was decided that people should work Monday through Friday, but there came the question as to when people should be allowed to retire. An oracle was consulted. After much deep thought the oracle jotted down something on a slip of paper, saying that if the number arrived at by the workers agreed with the number on the piece of paper, that number would indicate the age at which people should retire.

The workers were asked to write the days Monday through Friday on a blank calendar sheet, but in *any order* across the top and in *any different order* down the left side. A sample is shown in Figure 73. The days 1 through 25 were also filled in on the blank sheet, in the proper order.

After this was done the workers chose a particular day, say Thursday, and found the number at the intersection corresponding to the entry for Thursday in the vertical column and the horizontal row. In our example, the number would be 13, as shown in Figure 74.

This was done for each day of the work week. The result is shown in Figure 75. When the five circled numbers were added up, they totaled 65. This number exactly matched the number predicted by the oracle, and that's why, to this day, retirement age is 65.

There is no method other than the handling just described. Try it yourself with a square numbered from 1 to 25 as in Figure 73. Write the days of the week in any order across the top and in any different order down the left side. Then find where the Monday at the top and the Monday at the left intersect, the box where the Tuesday at the top and the Tuesday at the left intersect, and so on for each day. Total the five circled numbers and you will arrive at 65.

To present the trick, write the number 65 on a slip of paper, fold it and drop it into a drinking glass. Then write the numbers 1 to

	WED.	TUES.	THURS.	FRI.	MON.
FRI.	1	2	3	4	5
WED.	6	7	8	9	10
THURS.	11	12	13	14	15
TUES.	16	17	18	19	20
MON.	21	22	23	24	25

Fig. 73

	WED.	TUES.	THURS.	FRI.	MON.
FRI.	1	2		4	5
WED.	6	7		9	10
THURS.			⑬	14	15
TUES.	16	17	18	19	20
MON.	21	22	23	24	25

Fig. 74

	WED.	TUES.	THURS.	FRI.	MON.
FRI.	1	2	3	④	5
WED.	⑥	7	8	9	10
THURS.	11	12	⑬	14	15
TUES.	16	⑰	18	19	20
MON.	21	22	23	24	㉕

Fig. 75

25 in the square array. Have the spectator fill in the days of the week across the top and down the left side in any scrambled order. He then circles the numbers at the intersection points as described above. When he totals the five numbers he discovers that the total matches your prediction.

This is Martin Gardner's presentation for a matrix principle. Another application of the principle is described in "Auto Suggestion" (No. 84) in the next chapter.

GAMES AND BETS

Invariably people are attracted to games and betting propositions, especially where the outcome seems beyond any rational control. It is also true that simple number tricks take on unexpected appeal when presented as games.

Betting propositions serve other functions as well. For example, the dollar bill bet later on in this chapter would serve to introduce the idea of dollar-bill magic, allowing you to lead logically in to "Dollar-Bill Poker" (No. 5) in the first chapter or "Serial Secret" (No. 88) in the next chapter. The license-plate bet is a good introduction to "Auto Suggestion" (No. 84), the spectacular television-type lottery that ends this chapter, and so on.

74 MEXICAN BINGO

Required for this ingenious trick, the invention of Jack Avis, are a set of Mexican Bingo cards like those shown in Figure 76, and a deck of cards. The deck is given to the spectator at the beginning of the trick. The magician never touches the deck again.

You remark that the number cards are from the game of Mexican Bingo. In this game the numbers are not necessarily in order. The object of the game is to pick out the right numbers as quickly as possible.

Hand out the number cards, one to each of four spectators. Each spectator mentally decides on a number on his card.

The deck is removed from its case. The first spectator is asked to mentally spell out his thought-of number, one letter at a time, as he deals cards off the top of the deck. He deals a card for each letter. When he gets to the last letter, he is to stop.

By way of example, if he thought of the number One, he would deal a card as he mentally spells O, another card for N, and another

A

21	36	64	22
69	86	81	31
59	26	55	44
32	49	65	82
45	36		

B

68	67	29	72
24	89	71	85
76	84	25	39
43	47	35	34
53	57	58	63
48			

C

27	28	23
38	33	75
83	37	79
74	89	87

D

42	41	52
51	62	66
61	56	46

Fig. 76

card for E. As this explanation is given, the spectator deals cards off the top more or less as a practice exercise.

Once he understands this, have him go through the deal-spell procedure for the number he is thinking of. When he has dealt the last card, he takes the top card of the deck and places it in his pocket.

He hands the deck to the second spectator. This person has mentally chosen a number on his number card. He uses the same deal-spell procedure to arrive at a playing card and pockets this card. The same procedure is then followed by the third and fourth spectators.

You are now ready to test each spectator's aptitude at Mexican Bingo. The players remove their cards from their pockets and turn them face up. All four cards are Aces!

Remember that you never know what number was chosen by any spectator, nor do you touch the deck at any time. Although the numbers are chosen at random, the spectators show an uncanny ability at Mexican Bingo because they invariably choose the four Aces.

METHOD: Arrange the Aces at positions 13, 24, 36, and 45 from the top of the deck. That is the extent of the secret preparation.

The number cards must be filled out as in Figure 76. They can be made up on card stock and carried in the pocket or wallet.

Give card A to the first spectator, B to the second spectator, C to the third spectator, and D to the fourth spectator. Ask each to mentally choose a number on his card.

Remove the deck from its case. Either you or one of the spectators can deal cards as you demonstrate how the deal-spell process is to be carried out. Spell O-N-E aloud and deal a card for each letter. Then have the spectators deal-spell to their numbers. They must go in A, B, C, D order. The procedure is as already described, but be sure to remember that after each spectator spells out the chosen number, he takes the *next* card.

When the trick is performed as described above, the spectators will arrive at the four Aces. It is a surprising result and one that cannot be anticipated. The clever angle is that afterwards the spectators will recall counting to their numbers whereas they really spelled to the numbers.

75 A DOLLAR YOU CAN'T

The serial number on a dollar bill contains eight digits. Borrow a dollar from a spectator, fold it in half so the serial numbers are on the inside, and place it on the table. Bet a spectator that he can't name three of the digits on the bill. Oddly enough he will probably lose the bet.

76 A DOLLAR I CAN

Have the spectator remove a dollar bill from his pocket, fold it so the numbers are inside, and place the bill on the table. The bet begins the same way as "A Dollar You Can't," but this time you will guess the numbers on his bill.

Stare at the bill, then say, "Actually I remember spending this bill a week ago. Amazing that it traveled from town to town and ended up in your pocket. Let's see if I can win it back. As I recall, there is positively a 3 and a 7 on that bill. Care to bet on it?"

If he does, he will almost certainly lose. There is nothing crucial about the digits you call out. Call out any two digits and they will probably be on the bill.

77   THE FORTUNE 100

Bet a friend that he can quickly say 100 different words which do not contain the letters A, B, C, J, K, M, P, Q, or Z. Actually, anyone reading this book can rattle off a list of 100 such words instantly. All of the words are common English words. What words are they?

When the spectator gives up, just have him count from 1 to 100.

78   BAMBOOZLED

Each of the numbers from 1 to 10 can be spelled out. Thus, 1 is ONE, 2 is TWO, and so on. If the first ten numbers were spelled out, which three letters would appear the most often?

Surprisingly the answer is T-E-N. E appears the most frequently, N next most frequently, and T next.

Now for a more difficult question. Can you quickly determine which three letters appear most frequently when the numbers from 1 to 20 are written out? Again the answer is T-E-N.

We can put this information to use in an offbeat betting game. Have the spectator write out the numbers one after the other in a long chain, like this:

O N E T W O T H R E E F O U R . . .

He continues writing out the numbers through 20. Ask him to name a number between 1 and 120, then bet him that, if he counts the number of letters in the above chain, the count will end on a T, an E, or an N. If, for example, he calls out 11, and counts to the 11th letter, he will find that it is E, so you win. You will be right 55 percent of the time (assuming he calls out random numbers), which means that the game is in your favor.

79   RAPID BOXSCORES

Ask the spectator to guess last night's score in the Red-Blue hockey game (name two leading teams). When the spectator jots down his guess as to the score, you jot down under it the real score.

You have thus formed two two-digit numbers. Bet the spectator that you can instantly figure the product of the two numbers. If he doesn't believe you, you immediately jot down the product, and you are correct.

This is a little-known secret for performing rapid multiplication. Hockey scores tend to be in single digits, so the spectator is likely to jot down a two-digit number. Whatever number he writes, say to him, "You were close," and then you write the apparently correct score. Actually your first digit is the same as his first digit. The second digit you write is such that when added to the second digit of the spectator's number it will total 10. An example is:

$$58$$
$$\underline{52}$$

To calculate the product immediately, add 1 to the number on the upper left and multiply it by the number at the lower left. In our example 5 + 1 is 6, and 6 × 5 is 30. Jot down 30 as the first two digits of the product. Then multiply the number at the upper right by the number at the lower right and write this as the last two digits of the product. In the example the product of 8 × 2 is 16. The complete product is:

$$5\,8$$
$$\underline{5\,2}$$
$$3\,0\,1\,6$$

One more example should clear up any details. The spectator says the score in last night's game was 7-2. Always remark that he was close. Then tell him the score actually was 7-8. The product now will be obtained by multiplying 72 × 78. The spectator will want to do it with the traditional pencil and paper method. You use the system just described. Since the leftmost digit of his number is 7, add 1 to it to get 8. Multiply 8 by the leftmost digit of your number, 7, to get 56. Jot this down.

The spectator's rightmost digit is 2. Multiply this by your rightmost digit, 8, to get 16. The product of 72 and 78 is 5616.

80 PORTABLE POKER

This is a poker game that uses no cards. It can be played any time there is a pencil and paper handy. Go out of the room and have the spectator jot down five digits on a pad to represent a poker hand. He might choose the following hand:

$$4\,2\,5\,8\,3$$

Tell him to multiply this hand by 9 on the chance that it might improve the hand. He does, getting in this case the following result:

$$3\,8\,3\,2\,4\,7$$

Now tell him that he can add any pair to his hand. The pair must be between a pair of Aces and a pair of Nines. He might choose Eights. The poker hand is added to 88 as follows:

$$
\begin{array}{r}
3\,8\,3\,2\,4\,7 \\
+\quad\quad 8\,8 \\
\hline
3\,8\,3\,3\,3\,5
\end{array}
$$

He calls off this number to you, and you respond, "There are four Threes in the hand, but I think you chose to add, not Threes, but a pair of Eights—correct?"

Of course you are correct. As soon as he says so, you tell him his original poker hand!

METHOD: When the spectator reads off the digits in his final number, jot them down and add them together until they are reduced to a single digit. In our example you would arrive at

$$3 + 8 + 3 + 3 + 3 + 5 = 25$$

Reduce the number 25 to a single digit by adding the 2 to the 5. Your answer will be 7. Now add 9 to this number to see if you get a pair of identical digits. If you don't, add 9 again. The first time, $7 + 9 = 16$, which is not a matched pair of digits. Continue adding 9's until you get two identical digits. In our example you would keep adding 9's to get these totals:

$$25 \quad 34 \quad 43 \quad 52 \quad 61 \quad 70 \quad 79 \quad 88$$

Stop here because 88 is a matched pair. Thus you know that the spectator added a pair of Eights to his poker hand.

To determine his original poker hand, simply work backwards. Subtract 88 from the number he gave you (383,335 in our example) to arrive at 383,247. Then divide this number by 9, arriving at 42,583, his original poker hand.

The reason for doing the trick while you are in the next room is so that you can jot down the figures on a piece of paper and make quick calculations out of sight of the audience. The matching pair can be determined easily without pencil and paper. When you've reduced the spectator's number to a single digit, note whether it is odd or even. If it is even, take half of it and that is the value of the cards in the matching pair. If the digit is odd, add 9 and divide by 2 to find out the value of each card in the matching pair.

There is no real reason to make the above mental calculation, however. The trick is mystifying, especially if performed over the telephone, and the audience should be impressed when you tell them the value of the pair they added and then tell them the original poker hand.

81   THE GAME OF 21

In some parts of the world the game of blackjack or 21 is played a little differently. Six cards are arranged in the layout shown in Figure 77. Then the player is asked to connect cards to total 21.

The details are as follows. The spectator is to connect two numbers with a solid line such that the two numbers add exactly to 21. Then he is to connect two other numbers that add to 21. The line used to connect these two numbers must not cross the first line.

Finally, he is to connect the remaining two numbers if possible, with a line that does not cross either of the other lines. All lines must stay inside the layout.

Most spectators will find it easy to get two 21's, and impossible to get any more. But you can show them how easy it is to get three 21's while playing by the above rules.

METHOD: The solution suggested by Norm Osborn is shown in Figure 78. The reader can see that more pairs of cards can be added. A layout with four pairs is shown in Figure 79 along with the solution.

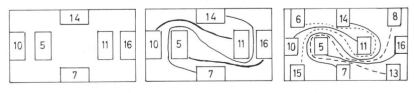

Fig. 77   Fig. 78   Fig. 79

82 TELEFOOLED

Telephone numbers can range from 0000 to 9999 (not counting exchanges), a total of 10,000 different numbers. Tell a spectator to open his telephone directory to any page and draw a circle around 20 consecutive numbers. Bet him that two of those numbers will have the same last two digits (i.e., 35 or 21 or 83). You don't say what two digits, only that the last two digits of two numbers will be the same.

The odds appear to be astronomically against you, but the truth is that you will win far more often than you will lose.

83 LICENSE-PLATE SHOWDOWN

The next time you're with a friend walking down the street, say to him, "When we round the corner onto the next block we'll pass at least ten parked cars. I'm willing to bet that one out of those first ten cars will have a double digit at the end, like 11 or 22. I'm so sure I'm willing to bet even money."

If he doesn't take the bait, point out that there are thousands of different license-plate numbers, but only ten double digits. Clearly the odds are weighed against you. In fact the only thing you have going for you is the confidence to believe that a double-digit license plate will actually be among the first ten cars.

At even money it seems like an irresistible proposition to your friend, but in fact the odds are steeply against him.

After your friend loses, you can bring up the subject of another stunt with license-plate numbers. When you reach your destination, you can spring the following routine on him.

84 AUTO SUGGESTION

The game here is a kind of televised lottery which can be built into a spectacular trick for the living room or stage. The complete presentation is as follows.

Recently a car manufacturer came up with the idea of awarding a new car to the person whose license plate most closely matched a set of random numbers printed in the manufacturer's national advertising. The problem of devising a truly random number was quickly realized, but eventually an airtight method was hit on.

Five major cities were chosen and listed. Next to each the committee listed car sales for typical weeks. This was done on five cards. The result was as shown in Figure 80.

On national television they arranged for a spectator to first satisfy himself that all numbers on all cards were different. Then that person mixed the cards and arranged them in any order he liked.

State maps with the five key locations were given to five different people. The maps looked like Figure 81. Each person could take any map he wanted. Individual participants could exchange maps if they wanted to.

The five people lined up in any order they chose. Each of them was given one of the number cards of Figure 80. Each man now held a map and a number card. To further insure that a purely random number would be chosen, each man could exchange his map or number card with those held by any other participant.

NEW YORK	12
DENVER	25
MIAMI	38
CHICAGO	51
DALLAS	64

NEW YORK	17
DENVER	30
MIAMI	43
CHICAGO	56
DALLAS	69

NEW YORK	35
DENVER	48
MIAMI	61
CHICAGO	74
DALLAS	87

NEW YORK	28
DENVER	41
MIAMI	54
CHICAGO	67
DALLAS	80

NEW YORK	5
DENVER	18
MIAMI	31
CHICAGO	44
DALLAS	57

Fig. 80

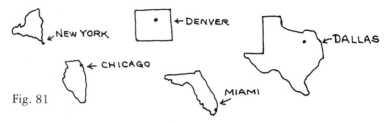

Fig. 81

Then each man noted which city he chose. Say the first man chose Denver. He looked at his number card and saw that the entry next to Denver was, for example, 25. This number was jotted down on a large blackboard.

The second man might have the Chicago map. On his number card the entry next to Chicago might be 56. This number was also jotted down.

The third man's map might have been Dallas. The entry next to Dallas on his number card might be 87. This number was noted on the blackboard.

The fourth man's map might have been Miami. On his number card the entry next to Miami might have been 54. This number was also noted.

Finally, the fifth man's map might have been New York. On his number card the entry next to New York might have been 5. This number was jotted down.

The five numbers on the blackboard are added up to produce the final random number. It might look like this:

$$
\begin{array}{r}
25 \\
56 \\
87 \\
54 \\
\underline{5} \\
227
\end{array}
$$

Up to this point the magician has used the above procedure to explain what was done on national television to arrive at a random license-plate number. But from the beginning a large sealed envelope has been in full view. When the random number 227 has been arrived at by the audience, the magician opens the envelope and removes a large license plate. "This is a copy of my license plate," he says. Then he turns the plate over to show the numbers and they exactly match the random number 227.

METHOD: As long as the above procedure is followed and each person chooses a different city, the numbers will always add to 227. The randomness of the number selection means nothing, but it looks so honest it suggests that it must be beyond the magician's control. This makes the outcome all that much more surprising.

Note too that if you totaled the numbers on any individual card, they would not sum to 227. It is only when a different entry on each card is taken that the force total of 227 is arrived at. Those who want an insight into why "Auto Suggestion" works might want to refer back to "Crazy Calendar" (No. 73) for a more basic exploitation of the principle.

The fact that all numbers are different, that number cards and maps may be distributed in any order, that the prediction is genuinely in view at all times from the start, makes this an immensely strong trick. Use large visible props and "Auto Suggestion" can become a sensational feature routine in any platform act.

GIANT MEMORY

The ability to memorize instantly long lists of words or numbers leads to spectacular feats of memory work. In this chapter we shall cover the simplest and best system of instant memorization, and then apply it to a number of ingenious tricks. It will take you approximately one evening to learn this system and it will be time well spent. With no special props or gimmicks, working either in the living room for friends or on a stage before a large audience, you will always be in a position to present staggering feats of mental prowess. Further, using the approach described here you have the option of presenting the effects as evidence of incredible memory or of sensational magical powers.

These claims will seem exaggerated until you learn the system and apply it to memory tricks and magical effects. Like the tricks of the lightning calculator, the feats of the memory expert are impressive merely because of the enormity of what is being attempted. That this system can be used for memory work and magical effects makes it doubly versatile.

85 THE MASTER MEMORY SYSTEM

Learning the basic system requires only that you associate each of the first ten numbers with key words. Later you will associate the key words with objects chosen by the audience. This is all there is to it. The system can be learned very quickly. The more often you use it, the better you will become at applying the word associations. Even if you miss, the audience will still be impressed at the gigantic feat you have attempted to perform.

Each number from 1 to 10 is associated with a word. The word for 1 is "gun." The word "gun" sounds the same as "one," so it is easy to make the association. In the same way, each of the other words rhymes with its associated number. Here is the basic word list:

One = Gun
Two = Shoe
Three = Tree
Four = Saw
Five = Hive
Six = Trix
Seven = Heaven
Eight = Gate
Nine = Mine
Ten = Pen

When you've learned the list, practice by placing five articles on the table. You are going to associate each of these articles with a number from 1 to 5. Say the articles from left to right are a coin, a pencil, a wallet, a watch, and a key.

The first article is the coin. Since it is first, you want to associate it with the key word "gun." Get a vivid picture in your mind that connects "gun" with "coin." Think of shooting a bullet through the coin the way a sharpshooter does. When the image is firmly in mind, go on to the next article.

In our example the next article is the pencil. Since it is the second article, we want to associate "shoe" with it. Again strive to get a vivid mental picture. Think of a shoe with a pencil stuck in the toe. When the image is clear, go on to the next article.

The third article is the wallet, and Three = Tree. Conjure up a picture of your wallet hanging from the highest branch of a tree. Then go on to the next article.

The watch is the fourth article and Four = Saw. This is easy. Picture someone sawing through a watch. Remember to conjure up a vivid image.

The fifth object is a key and Five = Hive. Think of a key unlocking a beehive.

Go back over this list to lock in the connections. You should then be able to say that the first object on the list is the coin. You know this because One = Gun and as soon as "gun" came to mind, a picture of a bullet whizzing through a coin should also have come to mind.

Two = Shoe. As soon as you think of "shoe" you should get a mental picture of a pencil sticking out of the toe of a shoe.

The same sort of approach is used for each of the other objects. If you want to recall the object that is fifth in the row, first recall that Five = Hive, then recall the mental picture you conjured up that connected hive with key.

Try the system with five objects as described above, then try it with five other objects and associate them with the digits 6 through 10. You should not practice for long hours. If you watch television, practice during commercials. Otherwise, devote five or ten minutes a day to the system. It can be learned in one evening but you have the option of practicing it a little at a time. In no time you will be able to memorize ten objects and their positions without effort. You will then be able to state with confidence which object is in sixth position or fourth position. You can even read off the objects forward or backward or in any order designated by the spectator.

When the spectator writes a list of ten objects on a sheet of paper or blackboard, associate the key words with the corresponding objects. Don't change the system once you've learned it because this will lead to confusion. If the first word he writes down is "truck," think of shooting at a truck. The image is vivid and easily recalled later on. Remember that regardless of the first word on his list, you will *always* think of shooting at that object with a gun.

The more active your imagination, the easier you will find it to memorize any list of objects. The association of key words with objects called out by the spectators can be ludicrous or absurd or merely silly. In each case you want an action picture—something as vivid as possible. The more striking the mental picture, the easier it will be to recall it later on.

Once you have the mental picture locked in, immediately go on to the next object on the spectator's list. Don't worry about recalling each and every object at every stage of the process. Once a word picture is mentally locked in, move to the next object on the list.

Remember that once the basic word list is committed to memory, it is unwise to change the system because there will be confusion when it comes to recalling key words. But if you learn the basic system and use just that system, you cannot go wrong.

86 INSTANT MEMORY

This is a basic memory routine using the Master Memory System and a good one to practice. Your audiences will be impressed even if you get one or two objects wrong. Following this routine we'll describe a subtle method that allows you to perform an amazing memory trick that would trouble a memory expert.

The routine is as follows. Someone jots down a list of ten objects. You look over the list, then turn your back. He calls out a number, say 4. You immediately tell him the fourth object on the list. He names 7 and you tell him the seventh object on the list. Then you call out the objects in numerical order, and follow by immediately calling them out in *reverse* order. The finish is something called "Mind Erasing," but we will defer that for the moment.

It is assumed that you have invested the small amount of time needed to memorize the Master Memory System. This means that if you were asked for the word associated with Three, you would immediately say "tree." Asked for the word associated with Nine, you should think of "mine." Four is "saw," Ten is "pen," and so on. We will now discuss the procedure used in typical memory demonstrations.

The spectator is asked to jot down a list of ten different objects. Tell him they should be objects that are easy to visualize. Have him write them on a large pad or blackboard. A sample list might look like this:

1. Truck
2. Cat
3. Pot
4. Chair
5. Hot Dog
6. Sweater
7. Television
8. Book
9. Bus
10. Newspaper

As the spectator writes each object, associate the objects with your list of key words. Thus, since Truck is 1 and One = Gun, you would think of shooting a bullet through a truck or shooting a gun at a truck.

You have a certain interval of time to memorize each object because it takes the spectator a few seconds to write down each word. Make the association with the first word while he is writing the second word. Make the association with the second word while he writes the third word, and so on. Then, one or two seconds after he writes the tenth word, you've got the entire list memorized. From the spectator's point of view you have memorized the list as soon as he finished writing it.

While the spectator writes the second word, you have created a mental picture that connects "truck" with "gun." When this has been done, you go on to create a vivid image that connects "cat" with "shoe," and another for "pot" and "tree." Continue in this way, connecting "chair" with "saw," "hot dog" with "hive," "sweater" with "trix," "television" with "heaven," "book" with "gate," "sun" with "mine," and "newspaper" with "pen." With practice you should be able to make the connections at a glance. Just remember to make the word picture as vivid as possible. It helps if the mental image is funny or absurd because then it is more easily recalled.

As soon as the spectator completes the list, turn your back. Walk away. Keep your back to him. Ask him to call out a number. As soon as he does, connect that number with your key word. If for example he says 2, connect Two with Shoe. As soon as you do, the word picture should spring to mind. In this case you would have connected "shoe" with "cat," so you know the second object on the spectator's list is "cat."

Go on to name the fourth object, the sixth, and so on. Then name the objects at odd-numbered positions. This may seem cold and matter-of-fact in print but it is a dramatic effect from the layman's point of view. He could not remember those ten objects in any order, let alone in precise numerical order, so what you are demonstrating looks impossible.

Finish by naming the objects in numerical order and then in reverse numerical order. Don't make it look too easy. When he names 7, for example, say, "I can't get it. Let me go back to that one." Pretend to have trouble with the seventh object right along. Then, for a dramatic finish, say, "Let's go back to that seventh object again. I think, in fact I'm sure it was, not stereo, not radio. Now I see it, television!"

You can see from this discussion how easy it is to call out objects in any order decided upon by the spectator. The procedure is always the same. First think of the numerical position of the object, then think of the key word for that number, and then recall to mind the word picture you conjured up for that key word.

Before setting up the demonstration it is a good idea to emphasize the difficulty of what you are going to attempt to do. The best approach is one suggested by J. N. Hofzinser. Everyone knows the Pledge of Allegiance by memory. But who (you ask your audience) could instantly name the tenth word in the Pledge? The twenty-third word? Give the audience a moment to think about it. Then describe what you are going to do. You are not going to memorize a list of random objects. You are going to attempt the far more difficult task of memorizing the objects in any order called out by the audience. To the average layman the feat demonstrates incredible powers of memory.

87 DOUBLE MASTER

When you feel you know the Master Memory System so well that it is no longer a challenge, you might be interested in learning an amusingly simple way to double the number of objects you can memorize.

The key ingredient is that the objects between 11 and 20 are mentally grouped in a second list. In this list the key words are exactly the same as in the list for the objects from 1 to 10, but there is a slight change. Group One follows the basic list of key words, but Group Two has associated with it the word Blue (Two = Blue). Thus every object between 11 and 20 is connected to a key word, but the object is thought of as being blue in color.

Here is a sample list you can practice with. Suppose the spectator drew up the following list of objects:

1.	Table	11.	Phone
2.	Bottle	12.	Dog
3.	Soap	13.	Chimney
4.	Radio	14.	Flower
5.	Stove	15.	Football
6.	Lion	16.	Knee
7.	Watch	17.	Plane
8.	Tomato	18.	Tractor
9.	Money	19.	Bike
10.	Shirt	20.	Snowball

Items 1 through 10 are handled as already described (shooting a gun at a table, putting a bottle in a shoe, soap hanging from a tree, etc.). Objects 11 through 20 are connected with the same key words, but you are to think of each object as being blue in color.

Thus you would mentally picture shooting a gun at a blue phone, putting a blue dog in a shoe, a blue chimney perched in a tree, and so on. The images are inherently amusing, a factor that aids in memorization.

This doubles the number of objects you can memorize without the need of learning new key words.

88 SERIAL SECRET

One reason for including a chapter on memory work in a book of number tricks is that many of the tricks done with numbers can be streamlined using memory methods. A sample is discussed here. The problem of instantly memorizing the serial number on a dollar bill was discussed in "Dollar-Bill Poker" in the opening chapter. That was in essence a bluff method. Here we will describe a method that allows you actually to memorize the serial number.

It should be mentioned that mnemonic systems have been created for the memorization of long chains of digits. The best such systems associate each digit with a consonant. Groups of two or three consonants are then mentally combined to form words. These systems lie outside the scope of the present book, but there is an easy way to memorize serial numbers that uses the Master Memory System.

The system is the same as the one already described. The slight difference is that Pen does not stand for Ten but rather for 0. With this in mind you can memorize an eight-digit serial number as follows.

Think of the first four digits as you would a telephone number, grouping them in two sets of two digits each. Thus 4893 would be remembered as 48, 93. Just think, "Forty-eight, ninety-three."

The next four digits use the key words in the basic word list. If these four digits are 6821, think "trix, gate, shoe, gun." Various mental images may come to mind to help connect these four words. Thus you might think of a tricky gate and a shoe armed with a gun. These will aid your later recall but are not strictly necessary because it is easy to remember four simple words.

This is the key to the instant memorization of the serial number. You use one system for the first four digits and a completely different system for the next four digits. The systems do not interfere or overlap, so there can be no mental confusion when it comes time to recall the digits. Further, with practice you can call off the digits forward or backward.

Finally, remember to note the letter at the beginning and the end of the serial number because almost certainly the spectator will ask you to recite these letters after you call off the digits in the serial number.

The presentation is then as follows. Borrow a dollar bill and fold it in half so that the serial number is on the outside. Hold the folded bill up so that the spectator can see the serial number on the upper right half of the bill. Ask him if he can commit this number to memory.

What most people do not know is that the same serial number is at the bottom of the left side of the bill. So, while the spectator struggles to memorize the number he sees, you have ample opportunity to memorize the number you see.

If it appears as if he just might be succeeding, confuse him by asking if he can memorize the number backwards. Usually he will give up at this point. Glance at the number he was looking at as if you are going to memorize the serial number right now. Nod your head and hand him the bill. Then turn your back and recite the serial number. Pause, then reel it off backwards.

When you gain facility with one serial number, try to memorize the serial number on two different bills. You will then be in a position to read back the first four digits on one bill, the last four on the other, and so on.

89 MIND ERASING

It is easy to remember a list of ten objects but difficult to forget them. As proof, consider the following test. The spectator jots down ten objects on a pad or blackboard. With your back turned you call off the objects in any order. Then you state that you will erase any nine of the ten objects from your mind.

The spectator crosses out items on his list, saying aloud, "I'm crossing out the fifth object, the second, the ninth, the third," and so on, in random order, until there is just one item left on the list. Instantly you name that object!

This type of test is seldom done by memory experts because it is not easy. But with the system described here you will do it the first time you try it. The approach is to work within the Master Memory System but *without* using memory techniques.

The secret is this. When the spectator tells you he is crossing out the fifth object, the second, the ninth, and so on, simply add these numbers together. When you've added together the nine numbers he calls out, whatever total you have, subtract it from 55. The result will be the number of the object he didn't cross out. Nothing could be easier. Thus a trick that a memory expert would struggle with is seen to be remarkably easy.

In our example, suppose the spectator said, "I'm crossing out the fifth object, the second, the ninth, the third, the first, the eighth, the fourth, the tenth, and the sixth. Which object is left?"

Add the above numbers silently. You will get a total of 48. Subtract from 55 and you arrive at 7. You know that the seventh object is left. Associate 7 with "heaven," then recall to mind the object you connected with "heaven." Reveal the object in dramatic fashion.

If you are familiar with the Double Master system described earlier, you can do Mind Erasing with two spectators. Each has a list of ten objects, each crosses out nine random objects on his list. You then name the remaining object on each list.

90   PSEUDO MEMORY

Memory experts sometimes resort to subtle tricks to give the impression of being able to recall extremely long chains of random digits. One of the best is a stunt that allows for the apparent memorization of a deck of cards.

Taking any borrowed, shuffled deck the expert takes the top card, calls off its value, then places the card on the bottom of the deck. He does this for 20 or 30 cards, skipping picture cards. As he calls off the cards, a spectator jots them down in order on a pad.

Now the deck is shuffled. This destroys the order of the cards, thus preventing the memory expert from resorting to trickery with the cards. The *only* record of the order of the cards is the list on the pad held by the spectator.

The memory expert never touches the list. He goes to a far corner of the room and immediately calls off the names of the cards!

METHOD: It looks like a feat of memory but it is all a fake. It's easy to remember 20-digit or 30-digit numbers, or even 50-digit numbers. Simply recall to mind your own telephone number, then a friend's number, then another friend's number. That alone produces a 21-digit number. Rattle off the three numbers one after the other and it seems like a random 21-digit number.

To apply it to the problem of memorizing the deck, take cards off the top one at a time and glance at them. Pretend to call out their values as you come to them but really call off the telephone numbers you already know. Continue for two or three telephone numbers. When you come to a "0" in the phone number call out "10" to the spectator. He jots down the numbers he hears. The long string of digits looks random because in a sense it is.

Have someone then shuffle the deck. Remark that this is to avoid trickery. Actually you have already used trickery. By shuffling the deck the spectator destroys the evidence of your trickery.

Pretend to concentrate. Walk to a far corner of the room. Then call out the numbers. Obviously since you know them as several different phone numbers, you can break into the list at different points and read off segments (really individual phone numbers) in no apparent order.

The system can be further scrambled by using your Social Security number, zip code, date of birth, weight, height, etc. When you decide upon a list, jot it down and make sure that individual digits are repeated no more than four times. This is because in a deck of cards there are only four of any given value. Once you have decided on a set of numbers, use only that set. The memory feat looks astounding but the real secret is a subtle trick that short-circuits the hard work.

91 MEMOREASE

In all of the preceding tricks in this chapter you have demonstrated amazing abilities at instant memorization. In this trick the tables are turned and a spectator temporarily becomes endowed with a photographic memory.

If you have just performed "Serial Secret" (No. 88), remark that anyone can remember long chains of numbers. To illustrate, you have a packet of cards, each of which contains a digit. The cards are shown one after the other to three different spectators. Two of the spectators fail to remember the order of the cards but one spectator finds he can recall every card in order and even in reverse order!

METHOD: The secret is J. K. Hartman's ingenious adaptation of an idea of the author's. Required are ten blank pieces of cardboard or slips of paper. Find out the phone number of one spectator. Write the digits in order, one digit on each slip of paper. Include the area code as the first three digits.

As additional preparation, type or print on each of three small pieces of paper the instruction of Figure 82.

Begin by announcing that you are going to endow one spectator temporarily with a photographic memory, and to prove it you are going to give a brief test. You will work with three spectators, No. One on your left, No. Two on your right, No. Three in the middle. Spectator Two is the one who you say will receive special powers. He is also the spectator whose phone number has been jotted down on the cards.

Turning to Spectator One, show him the faces of the cards one at a time, letting him see each for a second and then dealing them in a face-down pile on the table. What is at work here is this: the first spectator is really seeing the second spectator's phone number, but in reverse order. The digits probably wouldn't make sense to him anyway, unless he knew Two's phone number, but this way you insure that he sees a group of scrambled digits.

Pick up the packet and go through the same sequence with Spectator Two. Since the original order of the cards is now reversed, he will become aware that the numbers he sees are the digits of his own telephone number.

Pick up the packet and repeat for Spectator Three, who, like Spectator One, sees the numbers in incorrect order.

Hand the test papers of Figure 82 and pencils to the spectators and request that they complete the test. Spectators One and Three saw the digits in reverse order so the chain of digits will be meaningless. If you don't tell the spectators ahead of time exactly what to expect, and don't hand them the test papers until the very end, it is highly unlikely that Spectators One and Three will remember specific digits. For Spectator Two the situation is reversed. Since he is being quizzed on his own telephone number, the test is childishly simple.

```
WRITE DOWN THE NUMBER YOU SAW:

    FIFTH          _____

    SEVENTH        _____

    NINTH          _____
```

Fig. 82

Pick up the packet of number cards after all the spectators have completed their test papers. Count aloud, dealing the cards onto the table. When you reach the fifth card, ask each spectator to say what he wrote. Turn the card face up to show that Spectator Two was correct. Resume the count and do the same with the seventh card and finally the ninth card. Note that if one of the other spectators should have remembered a digit or two at the start, they are seeing the digits now in reverse order. Thus, if Spectator One happened to remember the fifth digit he saw, the fifth digit he sees now is not the same as the fifth digit he saw originally. In effect you have stacked the cards against Spectators One and Three.

At the beginning, when showing the number cards to Spectator Two, make sure that the other spectators don't get a glimpse of the numbers. It is for this reason that Spectator Two should be off to one side.

Mr. Hartman also suggests that letters could be used instead of numbers. For example, Spectators One and Three see NAGAER-DLANOR while Spectator Two sees RONALD REAGAN. To direct attention away from the fact that the order of the cards is being reversed after each deal, proceed as follows. Deal the cards for Spectator One, then hand him his test paper and a pencil. You have thus created a time delay here, directing attention away from the packet of cards. After dealing the cards for the second spectator, hand him his test paper. Then deal the cards for the third spectator and hand him his test paper.

92   THE ULTRA MIND

Up to this point the ability to recall long lists of objects has been presented as a memory feat. It is impressive done that way, but there is another approach which allows you to present the same demonstration as proof of baffling mind-reading abilities. The Master Memory System is flexible enough to accommodate mind-reading tricks, so if you know the system you can immediately perform tricks like this one.

Ten business cards or squares of cardboard are given to the spectator. On each card he draws a picture.

You turn your back. Direct the spectator to turn all of the cards face down and mix them thoroughly. Have him choose any card and look at the picture. When he has done this, have him place the

JOHN DOE
MIND READER

Fig. 83

card on the table, picture side down. You cover the card with a book, then place a sheet of paper on top of the book, and then, without asking a question, you proceed to draw a picture that duplicates the drawing on the hidden card!

The effect can be repeated with one or two more cards. In some cases you may be able to describe the type of object (for bottle you might have to settle for, "I see a glass object," by way of example) but in all cases you will be correct.

It is not necessary to place the drawing under a book. The spectator can place it in his pocket. You do not glimpse the picture. There are no carbon impressions, no confederates. You never touch any card and you never ask a single question. Mentalists might hesitate to perform a test of this kind because it is not easily handled by traditional approaches, but with the right insight it becomes remarkably easy.

METHOD: Each card is secretly marked on the back. It is difficult to mark blank cards in such a way that the mark is inconspicuous to others but plainly visible to you. A simple way around the problem is to have cards like those shown in Figure 83. Each card bears your name and the word "Mind Reader."

You will use ten cards. It happens that "Mind Reader" contains ten letters. On the first card put a scratch mark through the M. On the second card put a scratch mark through the I. On the third card put a scratch mark through the N. This is shown by the arrow in Figure 83. Continue in this way with each card, marking a different letter on each.

This is the only preparation. Hand the spectator the packet of cards and a pencil. On the blank side of the first card he draws an object. Tell him it should be something easily visualized, like a flower or a car.

You stand so you can see what he draws. Remember that the first card is secretly marked on the back so you can identify it as the first card. As soon as he has drawn the object, make the mental connection between this object and "Gun," the first word on the list of key words in the Master Memory System.

Hand him the next card. This card is secretly marked on the back so you can identify it as the second card. Whatever object he draws, create a vivid mental picture that connects this object and "Shoe" (the second word on your list of key words).

Continue in this way with each of the remaining cards. What you are doing is nothing more than having the spectator fill out a list of objects, exactly as you did in "Instant Memory," but instead of listing the objects on a pad, he is drawing pictures of them on individual cards. Note too that it takes more time to draw a picture than it takes to scribble a word on a list. This gives you more time to make the mental connections between your key words and his objects.

When he has filled out all the cards have him turn them face down and mix them so well that even he doesn't know which pictures are on which cards. Then tell him to choose a card and look at the picture.

Have him put the card on the table so the picture-side is down. As you cover it with a book, note the secret mark on the back. If, say, this mark indicates that it is the fourth card, associate 4 with "Saw," then bring to mind the object you connected with "Saw." This is the object he drew on the face of the card. Place a sheet of paper on top of the book and, with a display of great concentration, duplicate the object on the card.

Although a simple trick from your point of view, it has a remarkable effect on laymen. To see it from their point of view, consider how you might duplicate the same effect if you didn't know the above system. The fact that you have duplicated a picture without seeing the picture or asking any questions elevates a simple trick to the level of a master mystery.

Since the trick is so strong you are likely to be asked to repeat it. Don't repeat it exactly because the trick would look too simple. Instead, have another object chosen and hidden under the book. Don't reveal the object directly. Talk about it, scribble something on the paper, finally describe the class of objects that this picture would fit into. If for example he chose the picture of a truck, say that you get an impression of something with wheels. Then say, "I'm having trouble with that one. Pick another picture. We'll come back to that one later."

Have him pick another card. He places it under the book and you struggle through a revelation. If, say, it's a picture of a dog, say, "I think it's a pet of some kind. Yes, I see now that it's a dog."

Pause here for dramatic effect, then say, "But the strange thing is that I get a persistent image of a dog sitting inside a truck. It doesn't quite make sense."

The spectator will be happy to tell you that you've just revealed the picture he drew on the *other* card!

It is not necessary for you to glimpse the secret marks on the back of the chosen card. The spectator can hide the picture in his pocket. When you turn around and get a sheet of paper, note the marks on the backs of the remaining nine cards. Add these numbers and subtract from 55. The result tells you the number on the spectator's card. This method takes time since you must scan nine cards, but it is dramatic. Save it for just the right occasion and you will be able to present an impossible-seeming trick.

There are a number of angles you can work into the "Ultra Mind" trick. For example, have someone draw objects on the first five cards. Then have his wife draw objects on the remaining five cards. He draws pictures in red pencil, she draws pictures in green pencil. Now you are in a position to reveal who drew the picture and what color was used. Then you go on to duplicate the hidden picture itself. The cards can be sealed in envelopes that are secretly marked on the outside.

Another approach is one that gets away from drawings entirely. Use ten large drinking cups. Secretly mark the bottom of each cup. Have small objects dropped into each cup. Place the cups on the table. While you turn your back someone mixes the cups and then inverts each cup on the table.

When you return to the table a spectator points to any cup. You place your hand over the cup and dramatically reveal the object hidden under the cup. The system is the same as already described, but since nothing is written down, the trick seems like genuine mind reading.

93 DIGIT MEMORY

In this chapter we have surveyed some of the easiest and best applications of the Master Memory System. There are many others, limited only by the reader's ingenuity and imagination. For example, Sam Schwartz has devised an excellent method of performing "Mind Erasing" which requires no figuring. In this approach you stand with your back to the audience and place both hands against the chest so the fingertips touch the chest. Picture the fingers as representing the digits from 1 to 10. As the spectator calls out each number, raise the corresponding finger. After he has called out nine digits in any order, you will be left with one finger touching the chest. The number represented by that finger corresponds to the object on the pad—the one not crossed out. This method is so simple that it makes an easy trick even easier.

MORE NUMBER MAGIC

Throughout this book we have discussed methods of predicting the sum of numbers chosen by the spectator, but there is a startling variation where the spectator predicts the sum of four numbers chosen by the magician. That trick, "Sum Fantastic," is the opening trick in this chapter. It is typical of the methods described here in the sense that all of them use cleverly concealed mathematical principles to produce strong mind-reading tricks.

94   SUM FANTASTIC

In this trick you pick four numbers and the spectator accurately predicts their sum. The effect clearly seems impossible, yet there is an ingenious approach which allows you to perform the trick with ease.

The effect is this. Two spectators each call out a three-digit number. One number might be 145 and the other 683. Put them together to get the six-digit number 145,683. The magician turns around a pad he has been holding in full view from the start. Written on it are four numbers, and incredibly, when they are added the sum turns out to be 145,683, exactly matching the total predicted by the spectators!

METHOD: The basic system was suggested by Fred Barlow and Will Dexter for a lightning-calculation stunt. It is easily adapted to a magical effect as follows.

On a pad jot down these numbers, leaving a space between the top two numbers as indicated:

57,739

31,284
22,088

Ask the first spectator to call out a number between 120 and 200. Request that there be no zeros in the number. He might call out 145. Pretend to write 145 at the bottom of the pad. In fact you mentally deduct 1 from each digit, getting 034. Jot *this* number down in the blank space:

57,739
34,
31,284
22,088

Note that you dropped the zero on the left as you would do anyway. Now ask a second spectator for a three-digit number between 200 and 1000. Again request that there be no zeros in his number.

He might call out 683. Pretend to write this number at the bottom of the pad, but instead mentally deduct 1 from each digit, getting 572. Write this number down in the blank space next to the 34 like this:

57,739
34,572
31,284
22,088

Quickly jot down the two spectators' numbers—145 and 683—on the bottom of the pad. That's all you do. From your point of view the trick is over. Nothing could be easier, yet you have set up a trick that is in the miracle class. All that remains is for you to build up the effect with dramatic presentation.

Have another spectator stand. Tear off the second or third sheet of the pad. Give him this sheet, along with a pencil. Tell him to jot down the numbers just called out. Here you refer back to the pad for the numbers called out by the spectators. Say, "I want you to write 145,683, just like that, as a six-digit number. These are the numbers called out a moment ago."

The spectator writes 145,683. Then say, "Now that you have the number, I won't need it." Tear off the bottom of the sheet, crumple it and place it in your pocket. You've just gotten rid of that portion of the top sheet which contains the chosen six-digit number.

Now say, "Before I came here tonight I jotted down some numbers on a pad, hoping to project one of them to someone in the audience. These are the numbers." Turn the pad so that the figures are visible to the audience.

"As you can see, 145 and 683 were not among the numbers I wrote, so as yet the test is not a success. But I think something even more remarkable has happened. Neither of these gentlemen called out a number on the pad, but I suspect they called out the total of *all four numbers!*"

Hand the pad to the spectator with the pencil. Have him total the numbers. You can even give him a pocket calculator to make the task easier. He should be nonplussed to discover that two people in the audience have somehow predicted the sum of four random numbers.

95 PSYCHIC CARD READING

In this unique effect you reveal a card merely thought of by the spectator. He is given several cards from the top of the deck and is told to shuffle and deal them, face up, into three rows of equal size. If there are any cards left over, he is to return them to the top of the deck. Your back is turned and you do not see the cards. The layout might look like Figure 84.

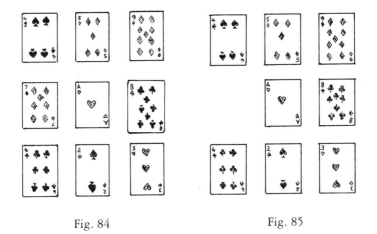

Fig. 84   Fig. 85

The spectator secretly pockets any card. Say he pockets the Seven of Diamonds. The layout now looks like Figure 85.

Have the spectator think of these cards as three rows of numbers. Tell him to add them. In this example the result will look like this:

$$
\begin{array}{r}
4\,5\,9 \\
1\,8 \\
\underline{6\,2\,3} \\
1\,1\,0\,0
\end{array}
$$

Tell the spectator to total all the digits in his answer and tell you the total. In this example the total is $1 + 1 + 0 + 0 = 2$. As soon as he announces the total, you tell him the card in his pocket.

METHOD: The cards on top of the deck are the Ace through Nine in the suits indicated in Figure 84. Note that red cards are odd-valued. It is easy to remember because "red" and "odd" have the same number of letters. Even-valued cards are black.

Remove these cards from the top. Hand them to the spectator and have him shuffle them. Instruct him to deal three rows with the same number of cards in each row. Suggest that he deal three cards in each row and continue doing this until he runs out of cards. By instructing him this way you make it seem that he has a random number of cards.

While you turn your back, he chooses and pockets any card. Then he notes the value of the remaining cards and adds them as described above. Treat the blank space as a 0. Whatever number he announces, subtract it from 9. If the number is greater than 9, subtract it from 18.

The result is the value of the spectator's card. Since odd-valued cards are red, when you know the value of the pocketed card, you also know its color. In the above example the chosen card is a 7. Since it is odd-valued, it must be red. If you can't recall the suit of the 7, take a guess and name either red suit. You might ask, "Was it a Heart?" In this example he will say no. You then say, "Then it must have been the Seven of Diamonds."

Suits have been arranged so that the two lowest red cards are Hearts and the two lowest black cards are Spades. The other red cards are Diamonds and the other blacks are Clubs. This is a simple system that is easily committed to memory.

Another example should clear up any questions. Suppose the spectator shuffles the cards and deals them out as in Figure 86. He pockets the Two of Spades. The layout will now look like Figure 87.

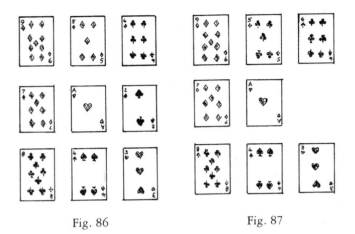

Fig. 86   Fig. 87

The spectator jots down the values, substituting a 0 for the blank space. The result in this example will be:

$$9\ 5\ 6$$
$$7\ 1\ 0$$
$$\underline{8\ 4\ 3}$$

He adds these numbers and gets 2509. You can have him add the digits in his answer, or you can have him call out the total. If he calls out 2509, add the digits, getting 16. Since you can't subtract 16 from 9, you subtract it from the next higher multiple of 9, in this case 18. The result is $18 - 16 = 2$. Knowing that he chose the Two, you also know it is black. Recalling that the two lowest blacks are Spades, you can then reveal that he chose the Two of Spades.

96 DEVIL'S DIAL

When in a restaurant ask that your guest jot down the last four digits of the telephone number of someone he knows. Explain that by a simple process you will convert the number into its psychic equivalent. Then, adding a local exchange, you will have him dial the psychic number.

When he dials the psychic phone number he asks the anonymous party on the other end to tell him his astrological sign. The other party might say, "Scorpio," and he's right!

METHOD: The system depends on an idea that appeared in *Mathematics Quarterly*. Begin by jotting down the last four digits of a friend's phone number. Subtract 19 from this number and remember the result. If the number is 7439, for example, subtract 19, getting 7420. This is the number you remember. This is done ahead of time and constitutes the only preparation.

When you're going to have lunch with an acquaintance, secretly find out his astrological sign. Then call your friend, the one whose phone number is 7439, and tell him the sign. Say it's Scorpio.

When you are in a restaurant, hand the acquaintance a pad and pencil. Turn your back. Have him jot down the last four digits of any telephone number. Say he writes 4312.

Have him multiply his number by 20, then subtract the original phone number, then divide this result by the original phone number. In our example, he would multiply 4312 by 20, getting 86,240. Then he subtracts the original number, getting 86,240 − 4312 = 81,928. Dividing 81,928 by 4312, he gets 19.

After he has done this, have him add 7420 to his result. Recall that 7420 is the four-digit number you got from the initial preparation. When he adds 7420 to 19, he gets 7439. But *this* is your friend's phone number.

Have him dial the number, adding the local exchange. Your friend answers, pretends to be puzzled at the call, then, when asked to name the acquaintance's sign, says "Scorpio," and hangs up.

Two points should be mentioned. If you called from the acquaintance's home phone he would spot the number when he got his next phone bill, if local calls are listed. Since he was so surprised that a random party guessed his sign, he might want to try it again. You don't want him to have the number, so you call from a public phone.

The other point is that the method described here can be used to force any number. Just deduct 19 from it and remember this result as your key. Go through the process described above, have the spectator add your key to his result, and you must arrive at the force number.

97 WHAT'S MY LINE?

This is a streamlined version of a trick marketed by Jack Miller and J. W. Sarles. The requirements are four envelopes and a match packet. Tear out the matches and toss them onto the table. Three envelopes are also placed on the table.

While you turn your back each spectator takes an envelope. There is a different occupation written on each envelope, so the spectators are encouraged to choose the job they'd like to have. Also on each envelope is the salary that the job pays. The spectators take their salary in matches and place the matches inside the envelopes. Then they hide the envelopes in their pockets.

You turn around and immediately reveal the job chosen by each spectator and what that job pays.

METHOD: The simplicity of this routine lies in the way the salary instructions are given to the spectators. The four envelopes are as shown in Figure 88. The Lawyer envelope has "Salary = 1" written on the outside. On the flap is written "Salary = 2." The flap is then tucked into the envelope. On a piece of paper that just fits the envelope is written, "Salary = 3." This piece of paper is then slid into the envelope. The Doctor and Banker envelopes are prepared as shown.

The Key List of Figure 89 is written on the fourth envelope and is not seen until the end of the trick. Keep it in your jacket pocket until then.

To present the trick, toss a full packet of 20 matches onto the table. You can use 20 toothpicks or 20 dollar bills in play money or 20 pennies. We'll assume that matches are used. Place the three envelopes on the table. Then turn your back.

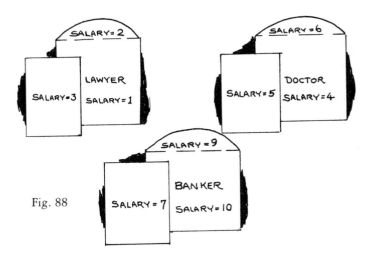

Fig. 88

KEY
LIST

1	=	B-10	L-3	D-6
3	=	B-10	D-5	L-2
4	=	D-4	L-3	B-9
5	=	L-1	D-5	B-9
6	=	L-1	B-7	D-6
7	=	D-4	B-7	L-2

Fig. 89

Ask each person to pick a job. It does not matter who chooses which envelope. Ask the first spectator to note the salary on the outside of his envelope, take that many matches and drop them into the envelope. Then he pockets the envelope.

Have the second spectator remove the slip of paper from his envelope. Written on this salary slip is the amount of his salary. Tell him to take that many matches, put them into the envelope, and pocket both the envelope and the salary slip.

The third spectator is to open the flap of his envelope and note the salary written on the flap. He takes that many matches, puts them in his envelope, and pockets the envelope.

Turn and face the spectators. Remove the envelope with the Key List from your pocket. Hold it so that only you can see the chart. Pick up the remaining matches and put them into the envelope. Secretly count the matches as you drop them in. The number of matches you put in the envelope tells you who has which envelope, his job, and his salary. If there is one match left when you turn around, the chart of Figure 89 tells you that 1 = B-10, L-3, D-6. This means that the first spectator is a banker and he has ten matches in his envelope, the second is a lawyer and he has three matches in his envelope, and the third is a doctor with six matches in his envelope.

Reveal this information in dramatic fashion.

To take another example, if there were five matches remaining, the Key List would tell you that 5 = L-1, D-5, B-9. This indicates that the first spectator is a lawyer with one match, the second is a doctor with five matches, and the third is a banker with nine matches.

There is another presentation angle which may be more puzzling. Explain to the audience that you know of a remarkable employment service that operates on supernatural wavelengths. Do the trick as described above, but the envelope you withdraw from your pocket contains a telephone number. Remove the slip of paper from the envelope and hand it to someone. Remark that they should dial the number and ask for Gail. They do, and she tells each spectator what job he has and what his salary is.

The key to the trick is the name you tell them to ask for. First, note how many matches are left after the spectators draw their salaries. Then remove the envelope of Figure 90 from the pocket. If there is one match remaining, have the spectator dial the number and ask for Alice. If there are three matches remaining, have the spectator ask for Carol. If four matches remain, he asks for Doris, and so on.

The medium on the other end hears the name called out and consults a chart like the one of Figure 89. If, say, the spectator asked for Carol, the medium would know that Carol = 3, and that 3 = B-10, L-3, D-6. From this she reveals who has what job and how much he's making.

To take another example, if the spectator asks for Gail, the medium knows that Gail = 7 according to the chart of Figure 90. Thus she knows that there are seven matches left. Consulting the Key List of Figure 89, she then knows that 7 = D-4, B-7, L-2. This tells her who has which job and how much he makes. She would reveal this information by saying, "The first spectator is a medical man and he draws four matches in pay. The second man is a banker who makes seven matches, and the third is a lawyer earning two matches."

Once the system becomes familiar to you, the presentation can be varied. The medium might want to ask nonsense questions like this: "What did you have for breakfast this morning? What color are your eyes? What's your favorite baseball team?" Pretending to mull over the spectator's answers, she then says, "This tells me you

1 = ALICE
3 = CAROL
4 = DORIS
5 = ETHEL
6 = FRAN
7 = GAIL

Fig. 90

are a doctor willing to work for four matches a week. At that salary we'll have no trouble placing you."

The jobs can also be varied to produce amusing situations. Thus one job can be banker and another bank robber and the third prosecuting attorney. Performed this way, the trick is an amusing and baffling demonstration of mind reading.

98 TRANSCENDENTAL BOOK TEST

An extremely clever method concealed behind an airtight presentation makes this one of the strongest mind-reading tricks around. There are no gimmicks or confederates, yet you are able to correctly reveal words chosen from a book in the spectator's possession from the start of the routine.

As the audience sees it, three books are used. The spectator chooses one for himself and you choose a book for yourself. The remaining book is isolated in another room.

You choose a page number in your book and have someone jot it down. Then the spectator chooses a page number and jots it down under yours. The numbers are added and from this total a random page number is arrived at. Someone goes into the next room, opens the book to the indicated page number, and concentrates on the words in the top line. Incredibly, you immediately reveal the words he sees!

Remember that you choose a number first. You don't influence the spectator's choice. In fact you never have to know his chosen number. The book in the next room is an ordinary book and you never go near it. Without asking a question you reveal the chosen words. These are impossible test conditions and you can make the most out of them when presenting the trick.

METHOD: The principle used in this trick was suggested by Richard Himber. Use any three books. Paperback mysteries are popular and usually available. Get three of them. Open one to page 43 and note the content of the top line. That is the only preparation. If you are in someone's home and want to do the trick, wait until you are alone in the room, then open a book, magazine, or newspaper to page 43 and note the top line. The preparation takes less than a minute.

From this point on, the book with the line you've memorized will be called the force book. That is because you will force this book in a subtle way. When ready to perform the "Transcendental Book Test," gather the three books, one of which is the force book. Place them on the table along with a pad and pencil.

Ask the assisting spectator to choose any one of the three books. He can change his mind as often as he likes. Stress that he may select *any* book. One of two things will happen. Either he will pick the force book or he won't. If he does, have him hide it in the next room. Then tell him to come back and choose one of the remaining books for himself.

In the more likely case the spectator will choose one of the other books. When he has done this, you pick up the other random book, leaving the force book on the table.

Say to the spectator, "You have a book and I have a book. Would someone hide the remaining book in the next room?"

Either way the force book ends up in the other room. Riffle the pages of your book and have the spectator say stop. Whatever page he stops you at, say, "We're going to leave the choice of page numbers entirely to chance. You've stopped me at page 57." Remember that whatever page he stops you at, tell him he stopped you at page 57. As you flip through the pages try to stop as close to page 57 as you can. When you announce that you were stopped at page 57, close the book. Have someone jot down your page number.

Tell the spectator, "Since my number is between 50 and 100, would you open your book to any page between 50 and 100? Don't tell me the page number. Write it down under my number."

Once the numbers have been recorded, have the spectator add them. To this point the arithmetic is as follows:

Your number	57
His number	71
	128

Say to the spectator with the pad and pencil, "Keep the total secret. If the total is over 100, disregard the first figure on the left. Take just the last two figures on the right and deduct that number from your number."

His number	71
Total	− 28
	43

Note in the above example that since the total is 128, the assistant will drop the 1, getting 28. He deducts this number from the spectator's page number to arrive at 43.

The spectator remembers this result, goes into the next room, and opens the book to this page number. Since page 43 is the page you looked at before the trick began, you are now able to reveal the wording of the first line on that page.

Another example would be this. The spectator calls out page 82. Your page number is always 57. The arithmetic looks like this:

$$
\begin{array}{lr}
\text{Your number} & 57 \\
\text{His number} & \underline{82} \\
& 139
\end{array}
$$

The 1 at the left is dropped off or crossed out. The resulting number is then subtracted from the spectator's number:

$$
\begin{array}{lr}
\text{His number} & 82 \\
\text{Total} & \underline{-39} \\
& 43
\end{array}
$$

Once again the result is 43, so the spectator would turn to page 43 in the force book and look at the top line. You then go on to reveal the words he looks at.

Note that the spectator has a free choice of books. He chooses a page number *after* you do, and you *never* know what number he chose. There are no gimmicks or confederates and you ask no questions. Play up these points and you will impress the audience with a spectacular mind-reading trick.

PREDICTING THE FUTURE

This chapter deals exclusively with the feat of predicting the sum of a group of numbers. The chapter opens with two easy methods that exploit the basic principle. The closing routine tightens up the conditions to produce a spectacular prediction.

99 SUPER PROPHET

The mentalist hands a die to each of three spectators. Each person calls out a number on his die. The first spectator might call out 6, the second spectator 4, and the third spectator 3. These digits are written as a three-digit number on a pad as 643.

When each spectator has chosen a digit, he crosses it out with a pencil. Then another three-digit number is formed using other digits on the three dice. The second number might be 415. This number is entered under 643.

The process continues until six three-digit numbers have been formed. Each digit on each die is used just once. The result might look like this:

$$
\begin{array}{r}
6\,4\,3 \\
4\,1\,5 \\
2\,5\,6 \\
1\,6\,2 \\
3\,3\,1 \\
\underline{5\,2\,4}
\end{array}
$$

When totaled, the result is seen to be 2331. The mentalist remarks that a week before the demonstration he mailed a sealed envelope to one of the people in the audience. That person stands, removes the envelope from his pocket, opens it, and reads aloud the following message: "The total will be 2331." The prediction thus exactly matches the total arrived at by the spectators.

METHOD: Simply follow the above instructions and the total will be 2331. Using dice limits the choice of digits from 1 to 6. This in turn makes it easy to add up the column of figures.

A week before the show, mail a letter to someone who you know will be present on the night when the trick is performed. The prediction says that the total will be 2331. Write in large, bold letters so that the prediction will be clearly visible when removed from the envelope and unfolded.

Try to obtain dice of different colors. Red, white, and green dice are usually available in stationery stores. They can also be found in board games requiring the use of dice. For greater visibility you might want to make up cardboard dice measuring about 6″ on a side.

Remember that each spectator must call out a different digit each time. This is really the secret to the trick since each column of numbers will contain the digits 1 through 6. Even though the order of the digits is random, the sum is fixed and unvarying. From this you can see why the trick must work.

"Super Prophet" is a streamlined and somewhat simplified version of "The Million-Dollar Prediction" (No. 101) at the end of this chapter. Although it is not quite as strong in terms of effect, it is easy to do. Because the total of 2331 is automatically produced, you can concentrate your attention on dramatic presentation.

100 ASTRO THOUGHT

Nine squares of cardboard are used. Each has a number on one side and a color on the other. The cards are placed number-side down on the table and mixed. Each spectator chooses three cards. Three numbers are formed from the cards. Although the numbers are random, their total is seen to have been correctly predicted by the magician.

METHOD: Make up nine cards to look like Figure 91. Three of the cards have large red dots or circles on the back. The numbers on the opposite sides are 8, 3, 4. The numbers are written in black pencil.

The second set of cards have large yellow dots on the back. The opposite sides have the numbers 9, 5, 1 written in black pencil.

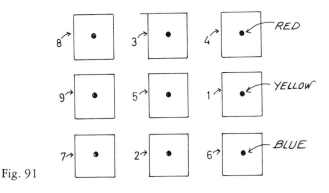

Fig. 91

The third set of cards have large blue dots on the back. The opposite sides have the numbers 7, 2, 6 written in black pencil.

On a large sheet of paper write, "The total will be 1665." Fold the paper and seal it in an envelope. If performing in the living room you can tape the prediction envelope to the door so that it is in plain view at all times.

Place the nine cards on the table, number-side down. Have a spectator mix the cards thoroughly. Tell him to pick any yellow card, any blue card, and any red card. After he has picked three cards, have a second spectator pick a card of each color. Then have a third spectator take the remaining three cards.

Pick up a large pad. Explain that the spectators can stand in any order. You then ask them to decide on an order for the three colors. It can be any order. Whatever they decide on, that is the order you will use for each spectator. This is an important point and one that will impress the audience.

Say the order they decide on is blue-yellow-red. Ask any spectator to stand and call out the number on his blue card. Say he calls out 2. Then ask for the number on his yellow card. It may be 5. Finally he calls out the number on his red card. It may be 4. The number you write is:

2 5 4

The second spectator now gives you the number on his blue card, then the number on his yellow card, then the number on his red card. These numbers might be 7, 9, 8. Write them below the first number as follows:

2 5 4
7 9 8

The third spectator might tell you that the number on his blue card is 6, the number on his yellow card is 1, and the number on his red card is 3. Write this three-digit number below the other two as follows:

$$2\ 5\ 4$$
$$7\ 9\ 8$$
$$\underline{6\ 1\ 3}$$

Have anyone add the three numbers. He will get a total of 1665. Then have him open the prediction envelope to check that the prediction matches the total.

It does not make any difference which order of colors is used. As long as the same order is used for each spectator, the total must be 1665. The reason why it works is that the numbering is done in accordance with the layout for a 3 × 3 magic square. Presented as described above, the mathematically predetermined outcome is neatly concealed.

101   THE MILLION-DOLLAR PREDICTION

This is the most impressive version of the number prediction. It uses no gimmicks or special apparatus. The trick may be performed for five people or 500. Since there is no advance preparation, "The Million-Dollar Prediction" can be done anywhere at a moment's notice.

Perhaps the most intriguing aspect of the prediction is that it deals solely with numbers suggested by the audience. Thus you do not predict something you could seemingly be able to control. Rather, you predict the total of four completely random numbers. The numbers are always different, their total is always different, yet the prediction is always correct.

As seen by the audience the effect is this. Each of four people calls out a four-digit number. The numbers are written on a large sheet of paper. The paper is then torn so that there is a digit on each of 16 pieces of paper. These pieces of paper are randomly mixed by the spectators themselves to form four different numbers which no one could possibly have known ahead of time. These random numbers are then added. Whatever the total, it exactly matches a total you predicted at the beginning of the trick.

You do not use sleight of hand or confederates. The numbers called out by the spectators are the numbers you use. No numbers are switched or miscalled. Most puzzling of all, at the start of the trick you have no idea what the prediction will be, yet as soon as you write it, you know it must be correct.

As with all impressive feats of mental magic, "The Million-Dollar Prediction" is based on a fundamentally simple premise. Because of this you can focus attention on presentation. Dramatically presented, "The Million-Dollar Prediction" is an astounding trick.

METHOD: Required are a pad of paper and a marking pen. The paper should be as large as possible so that the numbers can be written in large, bold strokes for greatest visibility. Ask for the assistance of four spectators from the audience. The prediction is so strong you may be suspected of using confederates. A simple way around this is to have on hand four Ping-Pong balls. Toss them out to different parts of the audience. Whoever catches a Ping-Pong ball is asked to stand and assist in the trick.

Ask each person to think of a four-digit number. The number can be the last four digits of a telephone number or part of a Social Security number or an important year in his life. The point to stress is that there are no limits. Each spectator can choose any four-digit number.

While the spectators think of numbers, tear off the top sheet of the pad. Fold it in half the long way, then in half again the long way. Make sure the creases are sharp because you will later tear the paper along the fold lines. Unfold the paper. It is now divided into four longitudinal sections thanks to the fold lines. Place the sheet flat on the pad.

Ask the first spectator for his number. Say it is 2719. Write the number across the top of the sheet of paper, one digit in each quarter, as shown in Figure 92. The dotted lines in Figure 92 represent the fold lines.

Proceed in the same way with each of the other spectator's numbers. Write the numbers as large as possible with the marking pen, one digit to each section of the paper. If the numbers are 2719, 8396, 4528, and 1783, the paper would be filled out as shown in Figure 92.

After each four-digit number is written, show the paper to the spectator and have him verify that the number you wrote is the number he called out. This gets around the suspicion that you may

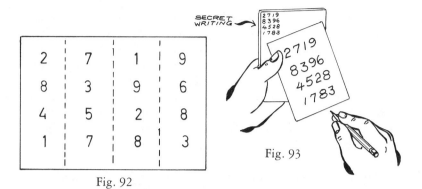

Fig. 92

Fig. 93

ignore the numbers actually called out or that you may alter the digits when you write them. Since each spectator positively sees that the number he chose is the number you wrote, the audience is convinced that all is honest. In fact the entire procedure is honest as you do not alter the numbers at any time.

What the audience is not aware of is that you secretly write each four-digit number on the pad itself. When each spectator calls out his number, jot it down on the pad as well as on the loose sheet, as indicated in Figure 93. When you show the writing on the loose sheet of paper to the audience, slide the loose sheet over the secret writing on the pad to hide the secret numbers from view.

This is the only secret work you do. If possible, jot down each number without looking directly at the pad. Have the spectator call out his number again, then write it on the loose sheet of paper. At the finish there will be two sets of figures. One set is on the loose sheet of paper. An identical set is written in tiny figures on the pad. The audience is aware only of the numbers on the loose sheet of paper.

If a slate is handy you can resort to a subtle method of hiding the secret writing in plain sight. Place the sheet of paper on the slate, using the slate as a backup, and record the numbers on the paper with pencil. Secretly jot them down on the slate as well. The pencil writing on the slate is visible to you but invisible to the audience. Later, when you write the prediction on the slate, write it in chalk directly over the penciled numbers. This obliterates the pencil writing from the audience's view even if members of the audience themselves handle the slate.

You are at the point where you have recorded four numbers on the loose sheet of paper. These same numbers have been secretly written on the pad. Place the pad on the table, writing-side down. Tear the loose sheet of paper into four strips as shown in Figure 94, tearing along the fold lines.

Remark that each strip of paper contains one digit from each spectator. By way of example, the first strip has one digit from the first spectator and three digits from the other spectators. Thus even if you could anticipate a four-digit number that someone might call out, his number would be only one-fourth of the total number of digits.

Tear the first strip into four pieces with one digit to each piece as shown in Figure 95. Give these four pieces of paper to the first spectator. Tear the second strip into four pieces and give them to the second spectator. Tear the third strip into four pieces and give them to the third spectator. Finally, tear the fourth strip into four pieces and give it to the fourth spectator. It is important that the four strips be distributed exactly as described here.

Now pick up the pad. Stand facing the audience. Tell each spectator to mix his four pieces of paper so that the digits are in scrambled order. Tell the first spectator to concentrate on the digits he holds. Pretend to get a telepathic impression from him. What you really do is add the rightmost column of digits on the pad. In this case you would add $9 + 6 + 8 + 3$ and get 26. Write down a large 6 on the bottom half of the sheet and make a mental note to carry the 2.

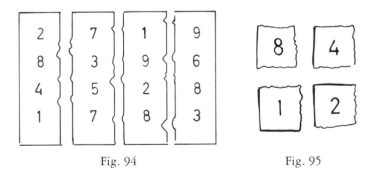

Fig. 94   Fig. 95

Now turn your attention to the second spectator. While he concentrates on his numbers, you pretend to get a telepathic image. In fact you add the column of numbers second from the right. In our example this column is $1 + 9 + 2 + 8$, which adds to 20. Recall that you carried 2 when you added up the first column, so the number is 22. Write a large 2 and mentally carry 2.

Repeat the process with each of the remaining spectators. It appears each time that you are getting a telepathic image, but in fact what you are doing is simply adding the four rows of numbers you secretly wrote on the pad.

In our example you would arrive at a sum of 17,426. This is the result you write on the pad. Tear off the lower half of the top sheet, that is, the portion containing the number 17,426, Figure 96, and place it back outwards in plain view. A simple way to accomplish this is to clip it to a spectator's jacket lapel with a bulldog clip. Of course the writing-side of the paper is hidden from the audience.

Tear off the remainder of the top sheet of the pad (the portion containing the secret writing), crumple it, and place it in your pocket. This gets rid of the incriminating evidence.

Turn to the first spectator. Ask him to call out any digit of the four he holds. Say he calls out the digit 4. Write this digit at the upper left corner of the new top sheet of the pad.

Have the second spectator call out any one of the digits he has. Say he calls out 5. Write this digit to the right of the 4. The third spectator might call out 9, chosen at random from the four digits (1, 9, 2, 8) he holds. Write the 9 to the right of the 5. The fourth spectator might decide to call out the digit 3 from among those he has. Write this to the right of 9. The result thus far is shown in Figure 97.

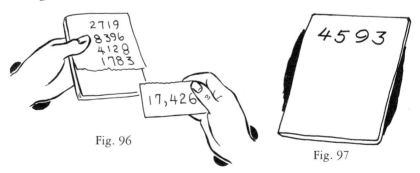

Fig. 96

Fig. 97

Now go back to the first spectator. Have him call out one of the remaining digits he holds. He might call out 8. Write the 8 under the 4. The second spectator might call out 3. Write the 3 under the 5. Proceed in this way to form the second number, which might be 8316.

Use the same process to form a third number and then a fourth number. In each case the spectators choose from among the digits in their possession. Since the four numbers are called out randomly, the new four-digit numbers that are formed bear no resemblance to the original numbers. Remember that each time you write a digit, turn the pad to the audience so they can see that the numbers you write are the numbers the spectators actually called out. The four numbers might look like Figure 98.

Fig. 98

Have someone total the four numbers. When he arrives at a total, have your prediction read aloud. As amazing as it will seem to the audience, your prediction will exactly match the total of the four numbers!

A CATALOG OF SELECTED
DOVER BOOKS
IN ALL FIELDS OF INTEREST

A CATALOG OF SELECTED
DOVER BOOKS
IN ALL FIELDS OF INTEREST

DRAWINGS OF REMBRANDT, edited by Seymour Slive. Updated Lippmann, Hofstede de Groot edition, with definitive scholarly apparatus. All portraits, biblical sketches, landscapes, nudes. Oriental figures, classical studies, together with selection of work by followers. 550 illustrations. Total of 630pp. 9⅛ × 12¼.
21485-0, 21486-9 Pa., Two-vol. set $29.90

GHOST AND HORROR STORIES OF AMBROSE BIERCE, Ambrose Bierce. 24 tales vividly imagined, strangely prophetic, and decades ahead of their time in technical skill: "The Damned Thing," "An Inhabitant of Carcosa," "The Eyes of the Panther," "Moxon's Master," and 20 more. 199pp. 5⅜ × 8½. 20767-6 Pa. $4.95

ETHICAL WRITINGS OF MAIMONIDES, Maimonides. Most significant ethical works of great medieval sage, newly translated for utmost precision, readability. Laws Concerning Character Traits, Eight Chapters, more. 192pp. 5⅜ × 8½.
24522-5 Pa. $5.95

THE EXPLORATION OF THE COLORADO RIVER AND ITS CANYONS, J. W. Powell. Full text of Powell's 1,000-mile expedition down the fabled Colorado in 1869. Superb account of terrain, geology, vegetation, Indians, famine, mutiny, treacherous rapids, mighty canyons, during exploration of last unknown part of continental U.S. 400pp. 5⅜ × 8½. 20094-9 Pa. $8.95

HISTORY OF PHILOSOPHY, Julián Marías. Clearest one-volume history on the market. Every major philosopher and dozens of others, to Existentialism and later. 505pp. 5⅜ × 8½. 21739-6 Pa. $9.95

ALL ABOUT LIGHTNING, Martin A. Uman. Highly readable nontechnical survey of nature and causes of lightning, thunderstorms, ball lightning, St. Elmo's Fire, much more. Illustrated. 192pp. 5⅜ × 8½. 25237-X Pa. $5.95

SAILING ALONE AROUND THE WORLD, Captain Joshua Slocum. First man to sail around the world, alone, in small boat. One of great feats of seamanship told in delightful manner. 67 illustrations. 294pp. 5⅜ × 8½. 20326-3 Pa. $4.95

LETTERS AND NOTES ON THE MANNERS, CUSTOMS AND CONDITIONS OF THE NORTH AMERICAN INDIANS, George Catlin. Classic account of life among Plains Indians: ceremonies, hunt, warfare, etc. 312 plates. 572pp. of text. 6⅛ × 9¼. 22118-0, 22119-9, Pa., Two-vol. set $17.90

THE SECRET LIFE OF SALVADOR DALÍ, Salvador Dalí. Outrageous but fascinating autobiography through Dalí's thirties with scores of drawings and sketches and 80 photographs. A must for lovers of 20th-century art. 432pp. 6½ × 9¼. (Available in U.S. only) 27454-3 Pa. $9.95

THE BOOK OF BEASTS: Being a Translation from a Latin Bestiary of the Twelfth Century, T. H. White. Wonderful catalog of real and fanciful beasts: manticore, griffin, phoenix, amphivius, jaculus, many more. White's witty erudite commentary on scientific, historical aspects enhances fascinating glimpse of medieval mind. Illustrated. 296pp. 5⅜ × 8¼. (Available in U.S. only)   24609-4 Pa. $7.95

FRANK LLOYD WRIGHT: Architecture and Nature with 160 Illustrations, Donald Hoffmann. Profusely illustrated study of influence of nature—especially prairie—on Wright's designs for Fallingwater, Robie House, Guggenheim Museum, other masterpieces. 96pp. 9¼ × 10¾.   25098-9 Pa. $8.95

LIMBERT ARTS AND CRAFTS FURNITURE: The Complete 1903 Catalog, Charles P. Limbert and Company. Rare catalog depicting 188 pieces of Mission-style furniture: fold-down tables and desks, bookcases, library and octagonal tables, chairs, more. Descriptive captions. 80pp. 9⅜ × 12¼.   27120-X Pa. $6.95

YEARS WITH FRANK LLOYD WRIGHT: Apprentice to Genius, Edgar Tafel. Insightful memoir by a former apprentice presents a revealing portrait of Wright the man, the inspired teacher, the greatest American architect. 372 black-and-white illustrations. Preface. Index. vi + 228pp. 8¼ × 11.   24801-1 Pa. $10.95

THE STORY OF KING ARTHUR AND HIS KNIGHTS, Howard Pyle. Enchanting version of King Arthur fable has delighted generations with imaginative narratives of exciting adventures and unforgettable illustrations by the author. 41 illustrations. xviii + 313pp. 6⅛ × 9¼.   21445-1 Pa. $6.95

THE GODS OF THE EGYPTIANS, E. A. Wallis Budge. Thorough coverage of numerous gods of ancient Egypt by foremost Egyptologist. Information on evolution of cults, rites and gods; the cult of Osiris; the Book of the Dead and its rites; the sacred animals and birds; Heaven and Hell; and more. 956pp. 6⅛ × 9¼.
22055-9, 22056-7 Pa., Two-vol. set $22.90

A THEOLOGICO-POLITICAL TREATISE, Benedict Spinoza. Also contains unfinished *Political Treatise*. Great classic on religious liberty, theory of government on common consent. R. Elwes translation. Total of 421pp. 5⅜ × 8½.
20249-6 Pa. $7.95

INCIDENTS OF TRAVEL IN CENTRAL AMERICA, CHIAPAS, AND YUCATAN, John L. Stephens. Almost single-handed discovery of Maya culture; exploration of ruined cities, monuments, temples; customs of Indians. 115 drawings. 892pp. 5⅜ × 8½.   22404-X, 22405-8 Pa., Two-vol. set $17.90

LOS CAPRICHOS, Francisco Goya. 80 plates of wild, grotesque monsters and caricatures. Prado manuscript included. 183pp. 6⅜ × 9⅜.   22384-1 Pa. $6.95

AUTOBIOGRAPHY: The Story of My Experiments with Truth, Mohandas K. Gandhi. Not hagiography, but Gandhi in his own words. Boyhood, legal studies, purification, the growth of the Satyagraha (nonviolent protest) movement. Critical, inspiring work of the man who freed India. 480pp. 5⅜ × 8½. (Available in U.S. only)
24593-4 Pa. $6.95

ILLUSTRATED DICTIONARY OF HISTORIC ARCHITECTURE, edited by Cyril M. Harris. Extraordinary compendium of clear, concise definitions for over 5,000 important architectural terms complemented by over 2,000 line drawings. Covers full spectrum of architecture from ancient ruins to 20th-century Modernism. Preface. 592pp. 7½ × 9⅜. 24444-X Pa. $15.95

THE NIGHT BEFORE CHRISTMAS, Clement C. Moore. Full text, and woodcuts from original 1848 book. Also critical, historical material. 19 illustrations. 40pp. 4⅝ × 6. 22797-9 Pa. $2.50

THE LESSON OF JAPANESE ARCHITECTURE: 165 Photographs, Jiro Harada. Memorable gallery of 165 photographs taken in the 1930s of exquisite Japanese homes of the well-to-do and historic buildings. 13 line diagrams. 192pp. 8⅜ × 11¼. 24778-3 Pa. $10.95

THE AUTOBIOGRAPHY OF CHARLES DARWIN AND SELECTED LETTERS, edited by Francis Darwin. The fascinating life of eccentric genius composed of an intimate memoir by Darwin (intended for his children); commentary by his son, Francis; hundreds of fragments from notebooks, journals, papers; and letters to and from Lyell, Hooker, Huxley, Wallace and Henslow. xi + 365pp. 5⅜ × 8. 20479-0 Pa. $6.95

WONDERS OF THE SKY: Observing Rainbows, Comets, Eclipses, the Stars and Other Phenomena, Fred Schaaf. Charming, easy-to-read poetic guide to all manner of celestial events visible to the naked eye. Mock suns, glories, Belt of Venus, more. Illustrated. 299pp. 5¼ × 8¼. 24402-4 Pa. $8.95

BURNHAM'S CELESTIAL HANDBOOK, Robert Burnham, Jr. Thorough guide to the stars beyond our solar system. Exhaustive treatment. Alphabetical by constellation: Andromeda to Cetus in Vol. 1; Chamaeleon to Orion in Vol. 2; and Pavo to Vulpecula in Vol. 3. Hundreds of illustrations. Index in Vol. 3. 2,000pp. 6⅛ × 9¼. 23567-X, 23568-8, 23673-0 Pa., Three-vol. set $41.85

STAR NAMES: Their Lore and Meaning, Richard Hinckley Allen. Fascinating history of names various cultures have given to constellations and literary and folkloristic uses that have been made of stars. Indexes to subjects. Arabic and Greek names. Biblical references. Bibliography. 563pp. 5⅜ × 8½. 21079-0 Pa. $9.95

THIRTY YEARS THAT SHOOK PHYSICS: The Story of Quantum Theory, George Gamow. Lucid, accessible introduction to influential theory of energy and matter. Careful explanations of Dirac's anti-particles, Bohr's model of the atom, much more. 12 plates. Numerous drawings. 240pp. 5⅜ × 8½. 24895-X Pa. $6.95

CHINESE DOMESTIC FURNITURE IN PHOTOGRAPHS AND MEASURED DRAWINGS, Gustav Ecke. A rare volume, now affordably priced for antique collectors, furniture buffs and art historians. Detailed review of styles ranging from early Shang to late Ming. Unabridged republication. 161 black-and-white drawings, photos. Total of 224pp. 8⅜ × 11¼. (Available in U.S. only) 25171-3 Pa. $14.95

VINCENT VAN GOGH: A Biography, Julius Meier-Graefe. Dynamic, penetrating study of artist's life, relationship with brother, Theo, painting techniques, travels, more. Readable, engrossing. 160pp. 5⅜ × 8½. (Available in U.S. only) 25253-1 Pa. $4.95

HOW TO WRITE, Gertrude Stein. Gertrude Stein claimed anyone could understand her unconventional writing—here are clues to help. Fascinating improvisations, language experiments, explanations illuminate Stein's craft and the art of writing. Total of 414pp. 4⅝ × 6⅜.   23144-5 Pa. $6.95

ADVENTURES AT SEA IN THE GREAT AGE OF SAIL: Five Firsthand Narratives, edited by Elliot Snow. Rare true accounts of exploration, whaling, shipwreck, fierce natives, trade, shipboard life, more. 33 illustrations. Introduction. 353pp. 5⅜ × 8½.   25177-2 Pa. $9.95

THE HERBAL OR GENERAL HISTORY OF PLANTS, John Gerard. Classic descriptions of about 2,850 plants—with over 2,700 illustrations—includes Latin and English names, physical descriptions, varieties, time and place of growth, more. 2,706 illustrations. xlv + 1,678pp. 8½ × 12¼.   23147-X Cloth. $89.95

DOROTHY AND THE WIZARD IN OZ, L. Frank Baum. Dorothy and the Wizard visit the center of the Earth, where people are vegetables, glass houses grow and Oz characters reappear. Classic sequel to *Wizard of Oz.* 256pp. 5⅜ × 8.

24714-7 Pa. $5.95

SONGS OF EXPERIENCE: Facsimile Reproduction with 26 Plates in Full Color, William Blake. This facsimile of Blake's original "Illuminated Book" reproduces 26 full-color plates from a rare 1826 edition. Includes "The Tyger," "London," "Holy Thursday," and other immortal poems. 26 color plates. Printed text of poems. 48pp. 5¼ × 7.   24636-1 Pa. $3.95

SONGS OF INNOCENCE, William Blake. The first and most popular of Blake's famous "Illuminated Books," in a facsimile edition reproducing all 31 brightly colored plates. Additional printed text of each poem. 64pp. 5¼ × 7.

22764-2 Pa. $3.95

PRECIOUS STONES, Max Bauer. Classic, thorough study of diamonds, rubies, emeralds, garnets, etc.: physical character, occurrence, properties, use, similar topics. 20 plates, 8 in color. 94 figures. 659pp. 6⅛ × 9¼.

21910-0, 21911-9 Pa., Two-vol. set $21.90

ENCYCLOPEDIA OF VICTORIAN NEEDLEWORK, S. F. A. Caulfeild and Blanche Saward. Full, precise descriptions of stitches, techniques for dozens of needlecrafts—most exhaustive reference of its kind. Over 800 figures. Total of 679pp. 8⅛ × 11.   22800-2, 22801-0 Pa., Two-vol. set $26.90

THE MARVELOUS LAND OF OZ, L. Frank Baum. Second Oz book, the Scarecrow and Tin Woodman are back with hero named Tip, Oz magic. 136 illustrations. 287pp. 5⅜ × 8½.   20692-0 Pa. $5.95

WILD FOWL DECOYS, Joel Barber. Basic book on the subject, by foremost authority and collector. Reveals history of decoy making and rigging, place in American culture, different kinds of decoys, how to make them, and how to use them. 140 plates. 156pp. 7⅞ × 10¾.   20011-6 Pa. $14.95

HISTORY OF LACE, Mrs. Bury Palliser. Definitive, profusely illustrated chronicle of lace from earliest times to late 19th century. Laces of Italy, Greece, England, France, Belgium, etc. Landmark of needlework scholarship. 266 illustrations. 672pp. 6¼ × 9¼.   24742-2 Pa. $16.95

ILLUSTRATED GUIDE TO SHAKER FURNITURE, Robert Meader. All furniture and appurtenances, with much on unknown local styles. 235 photos. 146pp. 9 × 12. 22819-3 Pa. $9.95

WHALE SHIPS AND WHALING: A Pictorial Survey, George Francis Dow. Over 200 vintage engravings, drawings, photographs of barks, brigs, cutters, other vessels. Also harpoons, lances, whaling guns, many other artifacts. Comprehensive text by foremost authority. 207 black-and-white illustrations. 288pp. 6 × 9. 24808-9 Pa. $9.95

THE BERTRAMS, Anthony Trollope. Powerful portrayal of blind self-will and thwarted ambition includes one of Trollope's most heartrending love stories. 497pp. 5⅜ × 8½. 25119-5 Pa. $9.95

ADVENTURES WITH A HAND LENS, Richard Headstrom. Clearly written guide to observing and studying flowers and grasses, fish scales, moth and insect wings, egg cases, buds, feathers, seeds, leaf scars, moss, molds, ferns, common crystals, etc.—all with an ordinary, inexpensive magnifying glass. 209 exact line drawings aid in your discoveries. 220pp. 5⅜ × 8½. 23330-8 Pa. $5.95

RODIN ON ART AND ARTISTS, Auguste Rodin. Great sculptor's candid, wide-ranging comments on meaning of art; great artists; relation of sculpture to poetry, painting, music; philosophy of life, more. 76 superb black-and-white illustrations of Rodin's sculpture, drawings and prints. 119pp. 8⅜ × 11¼. 24487-3 Pa. $7.95

FIFTY CLASSIC FRENCH FILMS, 1912–1982: A Pictorial Record, Anthony Slide. Memorable stills from Grand Illusion, Beauty and the Beast, Hiroshima, Mon Amour, many more. Credits, plot synopses, reviews, etc. 160pp. 8¼ × 11. 25256-6 Pa. $11.95

THE PRINCIPLES OF PSYCHOLOGY, William James. Famous long course complete, unabridged. Stream of thought, time perception, memory, experimental methods; great work decades ahead of its time. 94 figures. 1,391pp. 5⅜ × 8½. 20381-6, 20382-4 Pa., Two-vol. set $25.90

BODIES IN A BOOKSHOP, R. T. Campbell. Challenging mystery of blackmail and murder with ingenious plot and superbly drawn characters. In the best tradition of British suspense fiction. 192pp. 5⅜ × 8½. 24720-1 Pa. $5.95

CALLAS: Portrait of a Prima Donna, George Jellinek. Renowned commentator on the musical scene chronicles incredible career and life of the most controversial, fascinating, influential operatic personality of our time. 64 black-and-white photographs. 416pp. 5⅜ × 8¼. 25047-4 Pa. $8.95

GEOMETRY, RELATIVITY AND THE FOURTH DIMENSION, Rudolph Rucker. Exposition of fourth dimension, concepts of relativity as Flatland characters continue adventures. Popular, easily followed yet accurate, profound. 141 illustrations. 133pp. 5⅜ × 8½. 23400-2 Pa. $4.95

HOUSEHOLD STORIES BY THE BROTHERS GRIMM, with pictures by Walter Crane. 53 classic stories—Rumpelstiltskin, Rapunzel, Hansel and Gretel, the Fisherman and his Wife, Snow White, Tom Thumb, Sleeping Beauty, Cinderella, and so much more—lavishly illustrated with original 19th-century drawings. 114 illustrations. x + 269pp. 5⅜ × 8½. 21080-4 Pa. $4.95

SUNDIALS, Albert Waugh. Far and away the best, most thorough coverage of ideas, mathematics concerned, types, construction, adjusting anywhere. Over 100 illustrations. 230pp. 5⅜ × 8½.   22947-5 Pa. $5.95

PICTURE HISTORY OF THE NORMANDIE: With 190 Illustrations, Frank O. Braynard. Full story of legendary French ocean liner: Art Deco interiors, design innovations, furnishings, celebrities, maiden voyage, tragic fire, much more. Extensive text. 144pp. 8⅞ × 11¾.   25257-4 Pa. $11.95

THE FIRST AMERICAN COOKBOOK: A Facsimile of "American Cookery," 1796, Amelia Simmons. Facsimile of the first American-written cookbook published in the United States contains authentic recipes for colonial favorites—pumpkin pudding, winter squash pudding, spruce beer, Indian slapjacks, and more. Introductory Essay and Glossary of colonial cooking terms. 80pp. 5⅜ × 8½.
24710-4 Pa. $3.50

101 PUZZLES IN THOUGHT AND LOGIC, C. R. Wylie, Jr. Solve murders and robberies, find out which fishermen are liars, how a blind man could possibly identify a color—purely by your own reasoning! 107pp. 5⅜ × 8½. 20367-0 Pa. $2.95

ANCIENT EGYPTIAN MYTHS AND LEGENDS, Lewis Spence. Examines animism, totemism, fetishism, creation myths, deities, alchemy, art and magic, other topics. Over 50 illustrations. 432pp. 5⅜ × 8½.   26525-0 Pa. $8.95

ANTHROPOLOGY AND MODERN LIFE, Franz Boas. Great anthropologist's classic treatise on race and culture. Introduction by Ruth Bunzel. Only inexpensive paperback edition. 255pp. 5⅜ × 8½.   25245-0 Pa. $7.95

THE TALE OF PETER RABBIT, Beatrix Potter. The inimitable Peter's terrifying adventure in Mr. McGregor's garden, with all 27 wonderful, full-color Potter illustrations. 55pp. 4¼ × 5½.   22827-4 Pa. $1.75

THREE PROPHETIC SCIENCE FICTION NOVELS, H. G. Wells. *When the Sleeper Wakes, A Story of the Days to Come* and *The Time Machine* (full version). 335pp. 5⅜ × 8½. (Available in U.S. only)   20605-X Pa. $8.95

APICIUS COOKERY AND DINING IN IMPERIAL ROME, edited and translated by Joseph Dommers Vehling. Oldest known cookbook in existence offers readers a clear picture of what foods Romans ate, how they prepared them, etc. 49 illustrations. 301pp. 6⅛ × 9¼.   23563-7 Pa. $8.95

SHAKESPEARE LEXICON AND QUOTATION DICTIONARY, Alexander Schmidt. Full definitions, locations, shades of meaning of every word in plays and poems. More than 50,000 exact quotations. 1,485pp. 6½ × 9¼.
22726-X, 22727-8 Pa., Two-vol. set $31.90

THE WORLD'S GREAT SPEECHES, edited by Lewis Copeland and Lawrence W. Lamm. Vast collection of 278 speeches from Greeks to 1970. Powerful and effective models; unique look at history. 842pp. 5⅜ × 8½.   20468-5 Pa. $12.95

THE BLUE FAIRY BOOK, Andrew Lang. The first, most famous collection, with many familiar tales: Little Red Riding Hood, Aladdin and the Wonderful Lamp, Puss in Boots, Sleeping Beauty, Hansel and Gretel, Rumpelstiltskin; 37 in all. 138 illustrations. 390pp. 5⅜ × 8½. 21437-0 Pa. $6.95

THE STORY OF THE CHAMPIONS OF THE ROUND TABLE, Howard Pyle. Sir Launcelot, Sir Tristram and Sir Percival in spirited adventures of love and triumph retold in Pyle's inimitable style. 50 drawings, 31 full-page. xviii + 329pp. 6½ × 9¼. 21883-X Pa. $7.95

THE MYTHS OF THE NORTH AMERICAN INDIANS, Lewis Spence. Myths and legends of the Algonquins, Iroquois, Pawnees and Sioux with comprehensive historical and ethnological commentary. 36 illustrations. 5⅜ × 8½.
25967-6 Pa. $8.95

GREAT DINOSAUR HUNTERS AND THEIR DISCOVERIES, Edwin H. Colbert. Fascinating, lavishly illustrated chronicle of dinosaur research, 1820s to 1960. Achievements of Cope, Marsh, Brown, Buckland, Mantell, Huxley, many others. 384pp. 5¼ × 8¼. 24701-5 Pa. $8.95

THE TASTEMAKERS, Russell Lynes. Informal, illustrated social history of American taste 1850s–1950s. First popularized categories Highbrow, Lowbrow, Middlebrow. 129 illustrations. New (1979) afterword. 384pp. 6 × 9.
23993-4 Pa. $8.95

NORTH AMERICAN INDIAN LIFE: Customs and Traditions of 23 Tribes, Elsie Clews Parsons (ed.). 27 fictionalized essays by noted anthropologists examine religion, customs, government, additional facets of life among the Winnebago, Crow, Zuni, Eskimo, other tribes. 480pp. 6⅛ × 9¼. 27377-6 Pa. $10.95

AUTHENTIC VICTORIAN DECORATION AND ORNAMENTATION IN FULL COLOR: 46 Plates from "Studies in Design," Christopher Dresser. Superb full-color lithographs reproduced from rare original portfolio of a major Victorian designer. 48pp. 9¼ × 12¼. 25083-0 Pa. $7.95

PRIMITIVE ART, Franz Boas. Remains the best text ever prepared on subject, thoroughly discussing Indian, African, Asian, Australian, and, especially, Northern American primitive art. Over 950 illustrations show ceramics, masks, totem poles, weapons, textiles, paintings, much more. 376pp. 5⅜ × 8. 20025-6 Pa. $8.95

SIDELIGHTS ON RELATIVITY, Albert Einstein. Unabridged republication of two lectures delivered by the great physicist in 1920–21. *Ether and Relativity* and *Geometry and Experience*. Elegant ideas in nonmathematical form, accessible to intelligent layman. vi + 56pp. 5⅜ × 8½. 24511-X Pa. $3.95

THE WIT AND HUMOR OF OSCAR WILDE, edited by Alvin Redman. More than 1,000 ripostes, paradoxes, wisecracks: Work is the curse of the drinking classes, I can resist everything except temptation, etc. 258pp. 5⅜ × 8½. 20602-5 Pa. $4.95

ADVENTURES WITH A MICROSCOPE, Richard Headstrom. 59 adventures with clothing fibers, protozoa, ferns and lichens, roots and leaves, much more. 142 illustrations. 232pp. 5⅜ × 8½. 23471-1 Pa. $4.95

PLANTS OF THE BIBLE, Harold N. Moldenke and Alma L. Moldenke. Standard reference to all 230 plants mentioned in Scriptures. Latin name, biblical reference, uses, modern identity, much more. Unsurpassed encyclopedic resource for scholars, botanists, nature lovers, students of Bible. Bibliography. Indexes. 123 black-and-white illustrations. 384pp. 6 × 9. 25069-5 Pa. $9.95

FAMOUS AMERICAN WOMEN: A Biographical Dictionary from Colonial Times to the Present, Robert McHenry, ed. From Pocahontas to Rosa Parks, 1,035 distinguished American women documented in separate biographical entries. Accurate, up-to-date data, numerous categories, spans 400 years. Indices. 493pp. 6½ × 9¼. 24523-3 Pa. $11.95

THE FABULOUS INTERIORS OF THE GREAT OCEAN LINERS IN HISTORIC PHOTOGRAPHS, William H. Miller, Jr. Some 200 superb photographs capture exquisite interiors of world's great "floating palaces"—1890s to 1980s: *Titanic, Ile de France, Queen Elizabeth, United States, Europa,* more. Approx. 200 black-and-white photographs. Captions. Text. Introduction. 160pp. 8⅞ × 11¾. 24756-2 Pa. $10.95

THE GREAT LUXURY LINERS, 1927–1954: A Photographic Record, William H. Miller, Jr. Nostalgic tribute to heyday of ocean liners. 186 photos of *Ile de France, Normandie, Leviathan, Queen Elizabeth, United States,* many others. Interior and exterior views. Introduction. Captions. 160pp. 9 × 12. 24056-8 Pa. $12.95

A NATURAL HISTORY OF THE DUCKS, John Charles Phillips. Great landmark of ornithology offers complete detailed coverage of nearly 200 species and subspecies of ducks: gadwall, sheldrake, merganser, pintail, many more. 74 full-color plates, 102 black-and-white. Bibliography. Total of 1,920pp. 8⅜ × 11¼. 25141-1, 25142-X Cloth., Two-vol. set $100.00

THE COMPLETE "MASTERS OF THE POSTER": All 256 Color Plates from "Les Maîtres de l'Affiche", Stanley Appelbaum (ed.). The most famous compilation ever made of the art of the great age of the poster, featuring works by Chéret, Steinlen, Toulouse-Lautrec, nearly 100 other artists. One poster per page. 272pp. 9¼ × 12¼. 26309-6 Pa. $29.95

THE TEN BOOKS OF ARCHITECTURE: The 1755 Leoni Edition, Leon Battista Alberti. Rare classic helped introduce the glories of ancient architecture to the Renaissance. 68 black-and-white plates. 336pp. 8⅜ × 11¼. 25239-6 Pa. $14.95

MISS MACKENZIE, Anthony Trollope. Minor masterpieces by Victorian master unmasks many truths about life in 19th-century England. First inexpensive edition in years. 392pp. 5⅜ × 8½. 25201-9 Pa. $8.95

THE RIME OF THE ANCIENT MARINER, Gustave Doré, Samuel Taylor Coleridge. Dramatic engravings considered by many to be his greatest work. The terrifying space of the open sea, the storms and whirlpools of an unknown ocean, the ice of Antarctica, more—all rendered in a powerful, chilling manner. Full text. 38 plates. 77pp. 9¼ × 12. 22305-1 Pa. $4.95

THE EXPEDITIONS OF ZEBULON MONTGOMERY PIKE, Zebulon Montgomery Pike. Fascinating firsthand accounts (1805–6) of exploration of Mississippi River, Indian wars, capture by Spanish dragoons, much more. 1,088pp. 5⅜ × 8½. 25254-X, 25255-8 Pa., Two-vol. set $25.90

A CONCISE HISTORY OF PHOTOGRAPHY: Third Revised Edition, Helmut Gernsheim. Best one-volume history—camera obscura, photochemistry, daguerreotypes, evolution of cameras, film, more. Also artistic aspects—landscape, portraits, fine art, etc. 281 black-and-white photographs. 26 in color. 176pp. 8⅜ × 11¼.
25128-4 Pa. $14.95

THE DORÉ BIBLE ILLUSTRATIONS, Gustave Doré. 241 detailed plates from the Bible: the Creation scenes, Adam and Eve, Flood, Babylon, battle sequences, life of Jesus, etc. Each plate is accompanied by the verses from the King James version of the Bible. 241pp. 9 × 12.
23004-X Pa. $9.95

WANDERINGS IN WEST AFRICA, Richard F. Burton. Great Victorian scholar/adventurer's invaluable descriptions of African tribal rituals, fetishism, culture, art, much more. Fascinating 19th-century account. 624pp. 5⅜ × 8½. 26890-X Pa. $12.95

HISTORIC HOMES OF THE AMERICAN PRESIDENTS, Second Revised Edition, Irvin Haas. Guide to homes occupied by every president from Washington to Bush. Visiting hours, travel routes, more. 175 photos. 160pp. 8¼ × 11.
26751-2 Pa. $9.95

THE HISTORY OF THE LEWIS AND CLARK EXPEDITION, Meriwether Lewis and William Clark, edited by Elliott Coues. Classic edition of Lewis and Clark's day-by-day journals that later became the basis for U.S. claims to Oregon and the West. Accurate and invaluable geographical, botanical, biological, meteorological and anthropological material. Total of 1,508pp. 5⅜ × 8½.
21268-8, 21269-6, 21270-X Pa., Three-vol. set $29.85

LANGUAGE, TRUTH AND LOGIC, Alfred J. Ayer. Famous, clear introduction to Vienna, Cambridge schools of Logical Positivism. Role of philosophy, elimination of metaphysics, nature of analysis, etc. 160pp. 5⅜ × 8½. (Available in U.S. and Canada only)
20010-8 Pa. $3.95

MATHEMATICS FOR THE NONMATHEMATICIAN, Morris Kline. Detailed, college-level treatment of mathematics in cultural and historical context, with numerous exercises. For liberal arts students. Preface. Recommended Reading Lists. Tables. Index. Numerous black-and-white figures. xvi + 641pp. 5⅜ × 8½.
24823-2 Pa. $11.95

HANDBOOK OF PICTORIAL SYMBOLS, Rudolph Modley. 3,250 signs and symbols, many systems in full; official or heavy commercial use. Arranged by subject. Most in Pictorial Archive series. 143pp. 8⅛ × 11. 23357-X Pa. $8.95

INCIDENTS OF TRAVEL IN YUCATAN, John L. Stephens. Classic (1843) exploration of jungles of Yucatan, looking for evidences of Maya civilization. Travel adventures, Mexican and Indian culture, etc. Total of 669pp. 5⅜ × 8½.
20926-1, 20927-X Pa., Two-vol. set $13.90

DEGAS: An Intimate Portrait, Ambroise Vollard. Charming, anecdotal memoir by famous art dealer of one of the greatest 19th-century French painters. 14 black-and-white illustrations. Introduction by Harold L. Van Doren. 96pp. 5⅜ × 8½.
25131-4 Pa. $4.95

PERSONAL NARRATIVE OF A PILGRIMAGE TO AL-MADINAH AND MECCAH, Richard F. Burton. Great travel classic by remarkably colorful personality. Burton, disguised as a Moroccan, visited sacred shrines of Islam, narrowly escaping death. 47 illustrations. 959pp. 5⅜ × 8½.
21217-3, 21218-1 Pa., Two-vol. set $19.90

PHRASE AND WORD ORIGINS, A. H. Holt. Entertaining, reliable, modern study of more than 1,200 colorful words, phrases, origins and histories. Much unexpected information. 254pp. 5⅜ × 8½.
20758-7 Pa. $5.95

THE RED THUMB MARK, R. Austin Freeman. In this first Dr. Thorndyke case, the great scientific detective draws fascinating conclusions from the nature of a single fingerprint. Exciting story, authentic science. 320pp. 5⅜ × 8½. (Available in U.S. only)
25210-8 Pa. $6.95

AN EGYPTIAN HIEROGLYPHIC DICTIONARY, E. A. Wallis Budge. Monumental work containing about 25,000 words or terms that occur in texts ranging from 3000 B.C. to 600 A.D. Each entry consists of a transliteration of the word, the word in hieroglyphs, and the meaning in English. 1,314pp. 6⅜ × 10.
23615-3, 23616-1 Pa., Two-vol. set $35.90

THE COMPLEAT STRATEGYST: Being a Primer on the Theory of Games of Strategy, J. D. Williams. Highly entertaining classic describes, with many illustrated examples, how to select best strategies in conflict situations. Prefaces. Appendices. xvi + 268pp. 5⅜ × 8½.
25101-2 Pa. $7.95

THE ROAD TO OZ, L. Frank Baum. Dorothy meets the Shaggy Man, little Button-Bright and the Rainbow's beautiful daughter in this delightful trip to the magical Land of Oz. 272pp. 5⅜ × 8.
25208-6 Pa. $5.95

POINT AND LINE TO PLANE, Wassily Kandinsky. Seminal exposition of role of point, line, other elements in nonobjective painting. Essential to understanding 20th-century art. 127 illustrations. 192pp. 6½ × 9¼.
23808-3 Pa. $5.95

LADY ANNA, Anthony Trollope. Moving chronicle of Countess Lovel's bitter struggle to win for herself and daughter Anna their rightful rank and fortune—perhaps at cost of sanity itself. 384pp. 5⅜ × 8½.
24669-8 Pa. $8.95

EGYPTIAN MAGIC, E. A. Wallis Budge. Sums up all that is known about magic in Ancient Egypt: the role of magic in controlling the gods, powerful amulets that warded off evil spirits, scarabs of immortality, use of wax images, formulas and spells, the secret name, much more. 253pp. 5⅜ × 8½.
22681-6 Pa. $4.95

THE DANCE OF SIVA, Ananda Coomaraswamy. Preeminent authority unfolds the vast metaphysic of India: the revelation of her art, conception of the universe, social organization, etc. 27 reproductions of art masterpieces. 192pp. 5⅜ × 8½.
24817-8 Pa. $6.95

CHRISTMAS CUSTOMS AND TRADITIONS, Clement A. Miles. Origin, evolution, significance of religious, secular practices. Caroling, gifts, yule logs, much more. Full, scholarly yet fascinating; non-sectarian. 400pp. 5⅜ × 8½.
23354-5 Pa. $7.95

THE HUMAN FIGURE IN MOTION, Eadweard Muybridge. More than 4,500 stopped-action photos, in action series, showing undraped men, women, children jumping, lying down, throwing, sitting, wrestling, carrying, etc. 390pp. 7⅞ × 10⅝.
20204-6 Cloth. $24.95

THE MAN WHO WAS THURSDAY, Gilbert Keith Chesterton. Witty, fast-paced novel about a club of anarchists in turn-of-the-century London. Brilliant social, religious, philosophical speculations. 128pp. 5⅜ × 8½. 25121-7 Pa. $3.95

A CÉZANNE SKETCHBOOK: Figures, Portraits, Landscapes and Still Lifes, Paul Cézanne. Great artist experiments with tonal effects, light, mass, other qualities in over 100 drawings. A revealing view of developing master painter, precursor of Cubism. 102 black-and-white illustrations. 144pp. 8¾ × 6⅜. 24790-2 Pa. $6.95

AN ENCYCLOPEDIA OF BATTLES: Accounts of Over 1,560 Battles from 1479 B.C. to the Present, David Eggenberger. Presents essential details of every major battle in recorded history, from the first battle of Megiddo in 1479 B.C. to Grenada in 1984. List of Battle Maps. New Appendix covering the years 1967–1984. Index. 99 illustrations. 544pp. 6½ × 9¼. 24913-1 Pa. $14.95

AN ETYMOLOGICAL DICTIONARY OF MODERN ENGLISH, Ernest Weekley. Richest, fullest work, by foremost British lexicographer. Detailed word histories. Inexhaustible. Total of 856pp. 6½ × 9¼.
21873-2, 21874-0 Pa., Two-vol. set $19.90

WEBSTER'S AMERICAN MILITARY BIOGRAPHIES, edited by Robert McHenry. Over 1,000 figures who shaped 3 centuries of American military history. Detailed biographies of Nathan Hale, Douglas MacArthur, Mary Hallaren, others. Chronologies of engagements, more. Introduction. Addenda. 1,033 entries in alphabetical order. xi + 548pp. 6½ × 9¼. (Available in U.S. only)
24758-9 Pa. $13.95

LIFE IN ANCIENT EGYPT, Adolf Erman. Detailed older account, with much not in more recent books: domestic life, religion, magic, medicine, commerce, and whatever else needed for complete picture. Many illustrations. 597pp. 5⅜ × 8½.
22632-8 Pa. $9.95

HISTORIC COSTUME IN PICTURES, Braun & Schneider. Over 1,450 costumed figures shown, covering a wide variety of peoples: kings, emperors, nobles, priests, servants, soldiers, scholars, townsfolk, peasants, merchants, courtiers, cavaliers, and more. 256pp. 8⅜ × 11¼. 23150-X Pa. $9.95

THE NOTEBOOKS OF LEONARDO DA VINCI, edited by J. P. Richter. Extracts from manuscripts reveal great genius; on painting, sculpture, anatomy, sciences, geography, etc. Both Italian and English. 186 ms. pages reproduced, plus 500 additional drawings, including studies for *Last Supper, Sforza* monument, etc. 860pp. 7⅞ × 10¾. 22572-0, 22573-9 Pa., Two-vol. set $35.90

THE ART NOUVEAU STYLE BOOK OF ALPHONSE MUCHA: All 72 Plates from "Documents Decoratifs" in Original Color, Alphonse Mucha. Rare copyright-free design portfolio by high priest of Art Nouveau. Jewelry, wallpaper, stained glass, furniture, figure studies, plant and animal motifs, etc. Only complete one-volume edition. 80pp. 9⅜ × 12¼. 24044-4 Pa. $10.95

ANIMALS: 1,419 Copyright-Free Illustrations of Mammals, Birds, Fish, Insects, Etc., edited by Jim Harter. Clear wood engravings present, in extremely lifelike poses, over 1,000 species of animals. One of the most extensive pictorial sourcebooks of its kind. Captions. Index. 284pp. 9 × 12. 23766-4 Pa. $10.95

OBELISTS FLY HIGH, C. Daly King. Masterpiece of American detective fiction, long out of print, involves murder on a 1935 transcontinental flight—"a very thrilling story"—*NY Times*. Unabridged and unaltered republication of the edition published by William Collins Sons & Co. Ltd., London, 1935. 288pp. 5⅜ × 8½. (Available in U.S. only) 25036-9 Pa. $5.95

VICTORIAN AND EDWARDIAN FASHION: A Photographic Survey, Alison Gernsheim. First fashion history completely illustrated by contemporary photographs. Full text plus 235 photos, 1840–1914, in which many celebrities appear. 240pp. 6½ × 9¼. 24205-6 Pa. $8.95

THE ART OF THE FRENCH ILLUSTRATED BOOK, 1700–1914, Gordon N. Ray. Over 630 superb book illustrations by Fragonard, Delacroix, Daumier, Doré, Grandville, Manet, Mucha, Steinlen, Toulouse-Lautrec and many others. Preface. Introduction. 633 halftones. Indices of artists, authors & titles, binders and provenances. Appendices. Bibliography. 608pp. 8⅜ × 11¼. 25086-5 Pa. $24.95

THE WONDERFUL WIZARD OF OZ, L. Frank Baum. Facsimile in full color of America's finest children's classic. 143 illustrations by W. W. Denslow. 267pp. 5⅜ × 8½. 20691-2 Pa. $7.95

FOLLOWING THE EQUATOR: A Journey Around the World, Mark Twain. Great writer's 1897 account of circumnavigating the globe by steamship. Ironic humor, keen observations, vivid and fascinating descriptions of exotic places. 197 illustrations. 720pp. 5⅜ × 8½. 26113-1 Pa. $15.95

THE FRIENDLY STARS, Martha Evans Martin & Donald Howard Menzel. Classic text marshalls the stars together in an engaging, nontechnical survey, presenting them as sources of beauty in night sky. 23 illustrations. Foreword. 2 star charts. Index. 147pp. 5⅜ × 8½. 21099-5 Pa. $3.95

FADS AND FALLACIES IN THE NAME OF SCIENCE, Martin Gardner. Fair, witty appraisal of cranks, quacks, and quackeries of science and pseudoscience: hollow earth, Velikovsky, orgone energy, Dianetics, flying saucers, Bridey Murphy, food and medical fads, etc. Revised, expanded In the Name of Science. "A very able and even-tempered presentation."—*The New Yorker*. 363pp. 5⅜ × 8. 20394-8 Pa. $6.95

ANCIENT EGYPT: Its Culture and History, J. E. Manchip White. From predynastics through Ptolemies: society, history, political structure, religion, daily life, literature, cultural heritage. 48 plates. 217pp. 5⅜ × 8½. 22548-8 Pa. $5.95

SIR HARRY HOTSPUR OF HUMBLETHWAITE, Anthony Trollope. Incisive, unconventional psychological study of a conflict between a wealthy baronet, his idealistic daughter, and their scapegrace cousin. The 1870 novel in its first inexpensive edition in years. 250pp. 5⅜ × 8½.   24953-0 Pa. $6.95

LASERS AND HOLOGRAPHY, Winston E. Kock. Sound introduction to burgeoning field, expanded (1981) for second edition. Wave patterns, coherence, lasers, diffraction, zone plates, properties of holograms, recent advances. 84 illustrations. 160pp. 5⅜ × 8¼. (Except in United Kingdom)   24041-X Pa. $4.95

INTRODUCTION TO ARTIFICIAL INTELLIGENCE: Second, Enlarged Edition, Philip C. Jackson, Jr. Comprehensive survey of artificial intelligence—the study of how machines (computers) can be made to act intelligently. Includes introductory and advanced material. Extensive notes updating the main text. 132 black-and-white illustrations. 512pp. 5⅜ × 8½.   24864-X Pa. $10.95

HISTORY OF INDIAN AND INDONESIAN ART, Ananda K. Coomaraswamy. Over 400 illustrations illuminate classic study of Indian art from earliest Harappa finds to early 20th century. Provides philosophical, religious and social insights. 304pp. 6⅜ × 9⅜.   25005-9 Pa. $11.95

THE GOLEM, Gustav Meyrink. Most famous supernatural novel in modern European literature, set in Ghetto of Old Prague around 1890. Compelling story of mystical experiences, strange transformations, profound terror. 13 black-and-white illustrations. 224pp. 5⅜ × 8½.   25025-3 Pa. $7.95

PICTORIAL ENCYCLOPEDIA OF HISTORIC ARCHITECTURAL PLANS, DETAILS AND ELEMENTS: With 1,880 Line Drawings of Arches, Domes, Doorways, Facades, Gables, Windows, etc., John Theodore Haneman. Sourcebook of inspiration for architects, designers, others. Bibliography. Captions. 141pp. 9 × 12.
                     24605-1 Pa. $8.95

BENCHLEY LOST AND FOUND, Robert Benchley. Finest humor from early 30s, about pet peeves, child psychologists, post office and others. Mostly unavailable elsewhere. 73 illustrations by Peter Arno and others. 183pp. 5⅜ × 8½.
                     22410-4 Pa. $4.95

ERTÉ GRAPHICS, Erté. Collection of striking color graphics: *Seasons, Alphabet, Numerals, Aces* and *Precious Stones.* 50 plates, including 4 on covers. 48pp. 9⅜ × 12¼.
                     23580-7 Pa. $7.95

THE JOURNAL OF HENRY D. THOREAU, edited by Bradford Torrey, F. H. Allen. Complete reprinting of 14 volumes, 1837–61, over two million words; the sourcebooks for *Walden,* etc. Definitive. All original sketches, plus 75 photographs. 1,804pp. 8½ × 12¼.   20312-3, 20313-1 Cloth., Two-vol. set $130.00

CASTLES: Their Construction and History, Sidney Toy. Traces castle development from ancient roots. Nearly 200 photographs and drawings illustrate moats, keeps, baileys, many other features. Caernarvon, Dover Castles, Hadrian's Wall, Tower of London, dozens more. 256pp. 5⅜ × 8¼.   24898-4 Pa. $7.95

AMERICAN CLIPPER SHIPS: 1833–1858, Octavius T. Howe & Frederick C. Matthews. Fully-illustrated, encyclopedic review of 352 clipper ships from the period of America's greatest maritime supremacy. Introduction. 109 halftones. 5 black-and-white line illustrations. Index. Total of 928pp. 5⅜ × 8½.
25115-2, 25116-0 Pa., Two-vol. set $21.90

TOWARDS A NEW ARCHITECTURE, Le Corbusier. Pioneering manifesto by great architect, near legendary founder of "International School." Technical and aesthetic theories, views on industry, economics, relation of form to function, "mass-production spirit," much more. Profusely illustrated. Unabridged translation of 13th French edition. Introduction by Frederick Etchells. 320pp. 6⅛ × 9¼. (Available in U.S. only) 25023-7 Pa. $8.95

THE BOOK OF KELLS, edited by Blanche Cirker. Inexpensive collection of 32 full-color, full-page plates from the greatest illuminated manuscript of the Middle Ages, painstakingly reproduced from rare facsimile edition. Publisher's Note. Captions. 32pp. 9⅜ × 12¼. (Available in U.S. only) 24345-1 Pa. $5.95

BEST SCIENCE FICTION STORIES OF H. G. WELLS, H. G. Wells. Full novel The Invisible Man, plus 17 short stories: "The Crystal Egg," "Aepyornis Island," "The Strange Orchid," etc. 303pp. 5⅜ × 8½. (Available in U.S. only)
21531-8 Pa. $6.95

AMERICAN SAILING SHIPS: Their Plans and History, Charles G. Davis. Photos, construction details of schooners, frigates, clippers, other sailcraft of 18th to early 20th centuries—plus entertaining discourse on design, rigging, nautical lore, much more. 137 black-and-white illustrations. 240pp. 6⅛ × 9¼.
24658-2 Pa. $6.95

ENTERTAINING MATHEMATICAL PUZZLES, Martin Gardner. Selection of author's favorite conundrums involving arithmetic, money, speed, etc., with lively commentary. Complete solutions. 112pp. 5⅜ × 8½. 25211-6 Pa. $3.95

THE WILL TO BELIEVE, HUMAN IMMORTALITY, William James. Two books bound together. Effect of irrational on logical, and arguments for human immortality. 402pp. 5⅜ × 8½. 20291-7 Pa. $8.95

THE HAUNTED MONASTERY and THE CHINESE MAZE MURDERS, Robert Van Gulik. 2 full novels by Van Gulik continue adventures of Judge Dee and his companions. An evil Taoist monastery, seemingly supernatural events; overgrown topiary maze that hides strange crimes. Set in 7th-century China. 27 illustrations. 328pp. 5⅜ × 8½. 23502-5 Pa. $6.95

CELEBRATED CASES OF JUDGE DEE (DEE GOONG AN), translated by Robert Van Gulik. Authentic 18th-century Chinese detective novel; Dee and associates solve three interlocked cases. Led to Van Gulik's own stories with same characters. Extensive introduction. 9 illustrations. 237pp. 5⅜ × 8½.
23337-5 Pa. $5.95

Prices subject to change without notice.

Available at your book dealer or write for free catalog to Dept. GI, Dover Publications, Inc., 31 East 2nd St., Mineola, N.Y. 11501. Dover publishes more than 400 books each year on science, elementary and advanced mathematics, biology, music, art, literary history, social sciences and other areas.